Jan Grabowski

On the evolution of InAs thin films grown on the GaAs(001) surface

Jan Grabowski

On the evolution of InAs thin films grown on the GaAs(001) surface

A study using molecular beam epitaxy and scanning tunneling microscopy

Südwestdeutscher Verlag für Hochschulschriften

Impressum/Imprint (nur für Deutschland/only for Germany)
Bibliografische Information der Deutschen Nationalbibliothek: Die Deutsche Nationalbibliothek verzeichnet diese Publikation in der Deutschen Nationalbibliografie; detaillierte bibliografische Daten sind im Internet über http://dnb.d-nb.de abrufbar.
Alle in diesem Buch genannten Marken und Produktnamen unterliegen warenzeichen-, marken- oder patentrechtlichem Schutz bzw. sind Warenzeichen oder eingetragene Warenzeichen der jeweiligen Inhaber. Die Wiedergabe von Marken, Produktnamen, Gebrauchsnamen, Handelsnamen, Warenbezeichnungen u.s.w. in diesem Werk berechtigt auch ohne besondere Kennzeichnung nicht zu der Annahme, dass solche Namen im Sinne der Warenzeichen- und Markenschutzgesetzgebung als frei zu betrachten wären und daher von jedermann benutzt werden dürften.

Verlag: Südwestdeutscher Verlag für Hochschulschriften GmbH & Co. KG
Dudweiler Landstr. 99, 66123 Saarbrücken, Deutschland
Telefon +49 681 37 20 271-1, Telefax +49 681 37 20 271-0
Email: info@svh-verlag.de

Zugl.: Berlin, TU, Diss., 2010

Herstellung in Deutschland:
Schaltungsdienst Lange o.H.G., Berlin
Books on Demand GmbH, Norderstedt
Reha GmbH, Saarbrücken
Amazon Distribution GmbH, Leipzig
ISBN: 978-3-8381-2716-3

Imprint (only for USA, GB)
Bibliographic information published by the Deutsche Nationalbibliothek: The Deutsche Nationalbibliothek lists this publication in the Deutsche Nationalbibliografie; detailed bibliographic data are available in the Internet at http://dnb.d-nb.de.
Any brand names and product names mentioned in this book are subject to trademark, brand or patent protection and are trademarks or registered trademarks of their respective holders. The use of brand names, product names, common names, trade names, product descriptions etc. even without a particular marking in this works is in no way to be construed to mean that such names may be regarded as unrestricted in respect of trademark and brand protection legislation and could thus be used by anyone.

Publisher: Südwestdeutscher Verlag für Hochschulschriften GmbH & Co. KG
Dudweiler Landstr. 99, 66123 Saarbrücken, Germany
Phone +49 681 37 20 271-1, Fax +49 681 37 20 271-0
Email: info@svh-verlag.de

Printed in the U.S.A.
Printed in the U.K. by (see last page)
ISBN: 978-3-8381-2716-3

Copyright © 2011 by the author and Südwestdeutscher Verlag für Hochschulschriften GmbH & Co. KG and licensors
All rights reserved. Saarbrücken 2011

Der Beginn aller Wissenschaften ist das Erstaunen, dass die Dinge sind, wie sie sind.
Aristoteles

Zusammenfassung

Halbleiternanostrukturen sind zurzeit von großem Interesse für weitreichende Anwendungen in der Elektronik und Optoelektronik. Viele dieser Bauteile, insbesondere für die optische Datenübertragung im Bereich großer Wellenlängen, die von grundlegender Bedeutung in der modernen Kommunikationstechnik ist, basieren auf InAs/GaAs-Quantenpunktstrukturen (QD). Auch wenn die Eigenschaften der InAs/GaAs-QDs bereits intensiv untersucht wurden, so ist doch immer noch nur sehr wenig über die Benetzungsschicht (WL) bekannt. Im Rahmen dieser Arbeit wurde der Verlauf der Entstehung dieses InAs-WL im Detail untersucht.

Dazu wurden mittels Molekularstrahlepitaxie (MBE) dünne InAs-Schichten im Bereich einer Monolage (ML) auf die GaAs(0 0 1) Oberfläche aufgedampft und anschließend mit reflektiver hochenergetischer Elektronenbeugung (RHEED) und insbesondere mit der Rastertunnelmikroskopie (STM) untersucht. Die dünnen InAs Schichten wurden in den beiden typischen Wachstumsbereichen gewachsen, auf der GaAs-c(4×4) und der GaAs-β2(2×4) rekonstruierten Oberfläche, mit variabler Schichtdicke von Submonolagen mit 0,09 ML InAs bis zu 1,65 ML InAs, bei der die kritische Schichtdicke für das QD-Wachstum überschritten wird. Dabei wurden drei grundsätzliche Wachstumsphasen entdeckt.

Bei niedrigen InAs-Bedeckungen adsorbiert das Indium bevorzugt an energetisch günstigen Positionen auf der Oberfläche in Ansammlungen von durchschnittlich acht Indiumatomen. In den STM-Aufnahmen erscheinen diese Ansammlungen als deutliche helle Signaturen. Es werden ein Strukturentwicklungsmodell vorgestellt und die elektronischen Eigenschaften sowie die Gitterverspannung diskutiert.

Bei einer InAs-Bedeckung von 0,67 ML transformiert die ursprüngliche Oberfläche in eine (4×3) rekonstruierte In$_{2/3}$Ga$_{1/3}$As-ML und deren detaillierte Struktur und Verspannungseigenschaften werden aufgezeigt.

Weiter aufgedampftes InAs bildet dann eine zweite Lage auf der InGaAs-ML, gekennzeichnet durch eine typische zick-zack Anordnung von (2×4) rekonstruierten Einheitszellen, die eine abwechselnde α2/α2-m Konfiguration besitzen. Im Gegensatz zu den vorherigen Oberflächenrekonstruktionen, bei denen die strukturelle Gitterverspannung effizient abgebaut werden kann, staut sich in dieser zweiten (2×4) rekonstruierten Schicht mit dem weiteren Einbau von Indiumatomen eine kompressive Gitterverspannung an. Wenn diese zweite Schicht vervollständigt ist, beinhaltet der resultierende doppelschichtige WL eine Gesamtmenge von 1,42 ML InAs.

An diesem Punkt führt die angestaute Gitterverspannung zum Stranski-Krastanow (SK) Wachstumsübergang vom zweidimensionalen zum dreidimensionalen Wachstum, und weiteres aufgewachsenes InAs sammelt sich in typischen dreidimensionalen Inseln, den QDs. Darüber hinaus führt die Gitterverspannung im WL zur Verlagerung von Material aus dem WL in die QDs. Dieser Reifungsprozess, letztendlich auch auf Kosten von Teilen des WL, kann eingeschränkt werden, wenn das Substrat direkt nach dem Wachstum sehr schnell abgekühlt wird (quenching).

Abstract

Semiconductor nanostructures are currently of high interest for a wide variety of electronic and optoelectronic applications. A large number of devices, in particular for the optical data transmission in the long-wavelength range, essential in modern communication, are based on InAs/GaAs quantum dot (QD) structures. Though the properties of the InAs/GaAs QDs have been extensively studied, only little is known about the formation and structure of the wetting layer (WL) yet. In the present work, the pathway of the InAs WL evolution is studied in detail.

For this purpose, InAs thin films in the range of one monolayer (ML) are deposited on the GaAs(0 0 1) surface by molecular beam epitaxy (MBE) and studied by reflection high energy electron diffraction (RHEED) and in particular by scanning tunneling microscopy (STM). The InAs thin films are grown in both typical growth regimes, on the GaAs-c(4×4) and the GaAs-$\beta2(2\times4)$ reconstructed surface, in a variety of thicknesses starting from submonolayers with 0.09 ML of InAs up to 1.65 ML of InAs exceeding the critical thickness for QD growth. In principle, three growth stages are found.

At low InAs coverages, the indium adsorbs in agglomerations of typically eight In atoms at energetically preferable surface sites. In the STM images, the signatures of these In agglomerations appear with a clear bright contrast. A structural model for the initial formation of these signatures is presented, and its electronic and strain related properties are discussed.

At an InAs coverage of about 0.67 ML the initial surface transforms into a (4×3) reconstructed $In_{2/3}Ga_{1/3}As$ ML and the detailed structure and strain properties of this surface are unraveled.

On top of the InGaAs ML further deposited InAs forms a second layer, characterized by a typical zig-zag alignment of (2×4) reconstructed unit cells, with an alternating $\alpha2/\alpha2$-m configuration. In contrast to the previous surface reconstructions, where structural strain is sufficiently reduced, this second (2×4) reconstructed InAs layer accumulates unfavorable amounts of compressive strain from the InAs incorporation. With a fully evolved second layer the complete two-layer WL contains a total amount of 1.42 ML of InAs.

At this point, the accumulated amount of strain induces the Stranski-Krastanow (SK) growth transition from two-dimensional to three-dimensional growth, and further deposited InAs accumulates in typical three-dimensional islands, the QDs. Moreover, the unfavorable strain in the WL leads to a relocation of InAs material from the WL into the QDs. This QD ripening, eventually at the account of parts of the WL, can be reduced by rapid quenching of the substrate immediately after growth.

Contents

1. Introduction   15

Part I. Theoretical and experimental background

2. Semiconductor surfaces — properties and growth   21
   2.1. Gallium arsenide — a III/V semiconductor 21
   2.2. The electron counting rule 23
   2.3. The GaAs(0 0 1) surface 23
       2.3.1. The GaAs(0 0 1)-c(4×4) surface reconstruction 25
       2.3.2. The GaAs(0 0 1)-$\beta 2(2\times 4)$ surface reconstruction 26
   2.4. Semiconductor nanostructures 28
       2.4.1. Epitaxial growth 29
       2.4.2. Stranski-Krastanow growth of InAs/GaAs quantum dots 30

3. Molecular beam epitaxy (MBE)   31
   3.1. Ultra high vacuum (UHV) 31
   3.2. Evaporation 32
       3.2.1. Theory of evaporation 33
       3.2.2. Knudsen cells 34
   3.3. Material distribution 36
       3.3.1. Material flux 36
       3.3.2. Beam equivalent pressure 37
   3.4. Surface growth mechanisms 38
   3.5. Reflection high energy electron diffraction (RHEED) 39
       3.5.1. Formation of diffraction patterns 40
       3.5.2. Growth control 42

4. Scanning tunneling microscopy (STM)   45
   4.1. The STM principle 45
   4.2. Surface imaging 46
       4.2.1. Modes of operation 47
       4.2.2. Contrast mechanisms 48

5. Experimental setup   51
   5.1. The UHV chamber system setup 51
       5.1.1. The MBE setup 51
       5.1.2. The STM setup 54
       5.1.3. The setup for preparation and further analysis 55
       5.1.4. Sample storage and handling 56
   5.2. Preparation of the experimental setup 57
   5.3. Preparation of the STM tips 57
   5.4. Preparation of the sample substrate 58
   5.5. Calibration of the sample temperature 59

Part II. Results and discussion

6. Homoepitaxial growth on GaAs(0 0 1)   63
   6.1. Introduction 63
   6.2. Experimental details 64
       6.2.1. Sample preparation 64
       6.2.2. Determination of the growth rate 65
   6.3. The GaAs(0 0 1)-c(4×4) reconstructed surface 67
       6.3.1. Sample growth parameters 67
       6.3.2. STM results 67
       6.3.3. Summary 72
   6.4. The GaAs(0 0 1)-$\beta 2(2\times 4)$ reconstructed surface 72
       6.4.1. Sample growth parameters 72
       6.4.2. STM results 73
       6.4.3. Summary 75

7. InAs thin film growth on GaAs(0 0 1)-c(4×4)   77
   7.1. Introduction 77
   7.2. Experimental details 77
       7.2.1. Sample preparation 77
       7.2.2. Determination of the growth rate 78
   7.3. InAs thin films at submonolayer coverages 78
       7.3.1. Sample growth parameters 78
       7.3.2. STM results 78
       7.3.3. Discussion: Formation and structure of the InAs signatures 84
       7.3.4. Summary 88
   7.4. InAs thin films at coverages close to one monolayer 89
       7.4.1. Sample growth parameters 89

	7.4.2. STM results	89
	7.4.3. Discussion: Formation and structure of the $In_{2/3}Ga_{1/3}As$ monolayer .	96
	7.4.4. Discussion: Formation and structure of the InAs islands on top of the $In_{2/3}Ga_{1/3}As$ monolayer	100
	7.4.5. Summary	104
7.5.	InAs thin films during quantum dot growth	104
	7.5.1. Sample growth parameters	104
	7.5.2. STM results	104
	7.5.3. Discussion: The InGaAs/InAs wetting layer during quantum dot formation	109
	7.5.4. Summary	110
7.6.	Strain effects during the evolution of the InAs wetting layer	111
	7.6.1. Strain at the GaAs(0 0 1)-c(4×4) surface	111
	7.6.2. Strain related to the InAs signatures on the GaAs-c(4×4) surface . . .	112
	7.6.3. Strain at the $In_{2/3}Ga_{1/3}As$-(4×3) reconstructed monolayer	112
	7.6.4. Strain at the InAs-(2×4) reconstructed second layer	114
	7.6.5. Strain during quantum dot formation	115
	7.6.6. Summary	116

8. InAs thin film growth on GaAs(0 0 1)-$\beta 2(2\times 4)$ 117

    8.1. Introduction 117
    8.2. Experimental details 117
        8.2.1. Sample preparation 117
        8.2.2. Estimation of the growth rate 118
    8.3. InAs thin films on GaAs(0 0 1)-$\beta 2(2\times 4)$ 118
        8.3.1. Sample growth parameters 118
        8.3.2. STM results 119
        8.3.3. Discussion 124
        8.3.4. Summary 127

9. Conclusion   129

Appendix   133
    A. Data of the beam equivalent pressure (BEP) 135
    B. Determination of the deposited InAs coverage 139
    C. Sample heating data 143

List of Abbreviations   145

Bibliography   147

1. Introduction

In the 20th century, the ongoing industrialization with its growing need for advanced technology has become a strong motor for scientific progress in many fields. Yet, among all those fields, the fast evolving information and communication technology probably has branded this century most [1]. Its electronic devices have become so essential for everyday life, a world without seems hardly imaginable.

In retrospective, the introduction of the semiconductor-based transistor in 1948 [2] may be considered the starting point of this technological development and of the semiconductor industry. Semiconductors soon became the key material for electronic and optoelectronic applications, for their compounds are versatilely tunable in electronic and structural properties to meet many applicational needs [3].

Since then, there has been an ongoing progress to further improve the efficiency of such devices, while decreasing their size concurrently. Computers that once filled whole rooms now fit the palm of one's hand. Undoubtedly, this process of miniaturization eventually facilitated the broad acceptance of electronics in everyday life and fundamentally changed the way we live.

However, the ongoing ambition to further minimize the size of electronics has also created novel challenges to modern physics in recent decades [4]. On the nanoscale, quantum effects that were negligible before, now become dominating the electronic properties of such devices [5]. Lateral confinement of charge carriers causes the quantization of energy states, leading to a complete discretization when charge carriers are localized in all three dimensions [6,7]. In such a *quantum dot* (QD) the occupation of only certain energy levels is allowed, comparable to the electronic structure of a single atom [8].

Low-dimensional semiconductor structures (LDS) were predicted to be promising for novel applications, as low-threshold laser diodes [9–12], high-speed transistors [13], single photon detectors/emitters [14] or quantum memories [15,16]. An important step towards the systematic usage of quantum effects by manipulating matter on the nanoscale is marked by the discovery of self-organization mechanisms in epitaxial growth of lattice mismatched compound semiconductor material systems in 1985 [17].

Basic material for most optoelectronic nanostructure applications are III/V compound semiconductors. Their energy band gap can be tuned by adapting the chemical composition of the alloy [18,19] which allows, e.g., the adjustment of the light emission wavelength [20]. Likewise the lattice constant of the crystal structure is tunable to some extent [21,22]. Although many material compositions have been investigated for their structural and electronic properties, only some are yet used commercially. Among those, the *indium arsenide/gallium arsenide* (InAs/GaAs) material system is currently one of the technologically most important.

Low-dimensional InAs/GaAs nanostructures, e.g. QDs, are currently of high interest for the use in ultra-fast electronics [23], storage devices [24] and in particular in optoelectronic applications, especially for optical data transmission in the long-wavelength range [25–29]. Recently, also InAs/GaAs QD precursors, so-called *submonolayer quantum dots* have come into focus for the use in high power disk lasers [30], infrared photo detectors [31], as well as very fast light emitters for optical chip-to-chip connections [32–34].

Thus, the InAs/GaAs QD system has been studied extensively in the recent years. It is well established that the quantum dots form in the *Stranski - Krastanow* growth mode [8, 35, 36] and their size and shape have been investigated in detail [37–43]. However, detailed knowledge about the evolution of an InAs monolayer (ML) before the transition from two-dimensional growth to three-dimensional growth ($2D \rightarrow 3D$ *transition*) is still missing [36]. On the other hand, information on the atomic structure and the strain induced formation principles of the intermixed InGaAs wetting layer (WL) is fundamental for the detailed understanding of QD growth and even more for the formation of submonolayer QDs. Moreover, the atomic ordering in the WL creates a strain profile at the surface, affecting the adsorption of further deposited material and influencing its crystal structure and its optoelectronic properties [44].

In this work, the evolution of the InAs/GaAs wetting layer from the bare GaAs(0 0 1) surface to the critical thickness of QD formation is investigated on the atomic scale. Increasing amounts of InAs material were deposited on the GaAs(0 0 1) surface by *molecular beam epitaxy* (MBE) and studied *in-situ* by *reflection high electron diffraction* (RHEED) and in particular by *scanning tunneling microscopy* (STM).

In the first part of this thesis, a brief introduction on the material system and growth is given in Chapter 2. Chapters 3 and 4 focus on the experimental techniques MBE/RHEED and STM, respectively. Chapter 5 then provides an overview on the experimental setup and the sample preparation.

The second part of this thesis is dedicated to the presentation and interpretation of the experimental results. In Chapter 6 the principles of homoepitaxy are introduced and homoepitaxy on both the GaAs(0 0 1)-c(4×4) and the GaAs(0 0 1)-$\beta2(2\times4)$ surface is used to derive basic information on the growth parameters, such as the appropriate V/III ratio and the growth rate or on the calibration of the substrate temperature. In addition, STM images of these well-known surface reconstructions are used to calibrate the STM instrument. Chapter 7 presents the experimental findings on the InAs thin film evolution during InAs growth on the GaAs(0 0 1)-c(4×4) surface from very low submonolayer coverages to the critical thickness for QD growth. The discussion hereby mainly focuses on the formation processes and the structural properties of the different growth stages of the InAs thin film and on the effects of structural strain. Chapter 8 presents the experimental findings on the

InAs thin film evolution during growth on the GaAs(0 0 1)-$\beta 2(2\times 4)$ surface. It is shown that the principal growth mechanisms during the InAs thin film evolution are similar in both growth regimes. Finally, Chapter 9 summarizes the fundamental conclusions that could be drawn from the presented findings and gives an outlook on prospective experiments to unravel further aspects of InAs/GaAs growth.

Part I.

Theoretical and experimental background

2. Semiconductor surfaces — properties and growth

2.1. Gallium arsenide — a III/V semiconductor

Gallium arsenide (GaAs) is a binary semiconductor, and similarly to most III-V materials it crystallizes in the zincblende crystal structure. As the chemical bond between arsenic and gallium atoms is of both ionic and covalent nature, the covalence leads to the formation of sp^3-hybridized electron orbitals of the bulk atoms. This sp^3-hybridization is responsible for the tetrahedral alignment of the bulk atoms, characteristic for the zincblende crystal structure. Each atom is thereby equidistantly surrounded by four atoms of the other kind as illustrated in Fig. 2.1 [45, 46]. The energies of the hybridized orbitals from As and Ga atoms are located in the *valence band* (VB) and the *conduction band* (CB) of the semiconductor, respectively [47].

If the symmetry of the crystal structure is reduced, e.g. at the surface where the crystal is no longer infinite, the hybridized orbitals cannot form bonds and will remain unsaturated (*dangling bonds*). This surface structure is commonly referred to as *bulk-truncated*. These

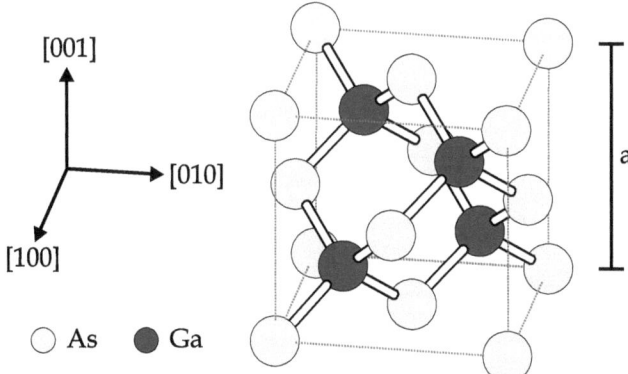

Figure 2.1: *The GaAs zincblende bulk crystal structure. The lattice parameter $a = 0.565$ nm [48] is defined as the side length of the cubic unit cell. Gallium atoms are illustrated in blue and arsenic atoms in light gray.*

dangling bonds are partly filled with electrons and thus cause additional surface states, which increase the surface energy of the semiconductor [49].

On the GaAs surface this bulk-truncated surface energy can be lowered by an electron transfer from the Ga dangling bonds in the CB to the As dangling bonds in the VB. With a depleted CB and a completely filled VB the semiconductivity at the surface is restored.

Naturally, such an electron transfer is only trivial on stoichiometric non-polar surfaces, where the number of group-III and group-V atoms is equal, e.g. the GaAs(1 1 0) oriented surface [50–52]. Surface Ga atoms would swap from sp^3-hybridization to sp^2-hybridization providing the excess electrons to fill the dangling bonds of the As atoms, allowing them to remain in sp^3-hybridization. The now sp^2-hybridized Ga atoms finally cause a change in the crystal surface geometry leading to a relaxation of Ga and As atoms, the so-called *buckling*, as illustrated in Fig. 2.2 [53–55].

Most low-index crystal surfaces, including the GaAs(0 0 1)-surface that was used for the investigations in this work, do not fulfill the stoichiometry criteria for such a simple relaxation and therefore undergo rearrangement processes to form stable reconstructed surfaces that will regain the semiconductivity at the surface as well as minimize the number of dangling bonds. The reconstructed surface thereby usually is the one with least free energy in the kinematically available range of possible reconstructions (see, e.g., [56]). Yet, as this kinematically available range is highly affected by the ambient conditions such as temperature and vapor pressure, different reconstructions for the same bulk-truncated surface are possible [57].

High-index bulk-truncated surfaces usually transform into reconstructions of small areas of low-index surfaces (*facets*) or into the terraced reconstruction of a low-index surface with similar orientation (*vicinal surfaces*). On the other hand, stable high-index GaAs surfaces, such as GaAs(1 1 4) [58,59] and GaAs(2 5 11) [60,61], have also been found.

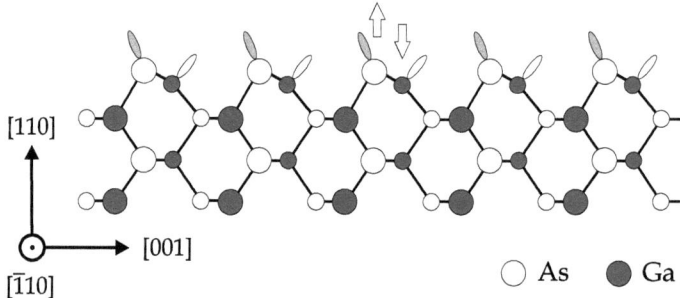

Figure 2.2: *Side view of the relaxed GaAs(1 1 0) surface. The hybridization transition of the uppermost Ga atoms from sp^3 to sp^2 leads to a retraction of these Ga atoms under the surface plane and thus an outstretching of neighboring surface As atoms. White arrows indicate this so-called buckling. The corresponding filled As dangling bonds are illustrated by gray ovals, empty Ga dangling bonds by white ovals. Smaller circles depict atoms located lower than the figure plane.*

2.2. The electron counting rule

The complete annihilation of surface charges — caused by dangling bonds on the bulk-truncated surface — is an appropriate criterion for the possible existence of stable surfaces and reconstructions, resulting in the principle of the *electron counting rule* (ECR). The ECR requires the number of available surface state electrons (i.e. from states that are not of bulk nature) to exactly fill all dangling bond states in the VB, leaving those in the CB completely depleted, in order to restore the surface semiconductivity [47, 55, 62, 63].

For III-V semiconductors in zincblende structure, such as GaAs, each atom has covalent bonds to four tetrahedrally aligned atoms of the other kind. Each bond is filled with two electrons of which gallium, as the trivalent part, contributes 3/4 electrons and arsenic, as the pentavalent part, 5/4 electrons.

As an example, on the GaAs(1 1 0) surface, a filled As dangling bond requires two electrons and the Ga dangling bond has to be empty. The surface stoichiometry delivers 3/4 electrons from Ga and 5/4 electrons from As, so that the electron transfer during surface relaxation [49, 55] will fulfill the ECR (Eq. 2.1):

$$\begin{aligned} \text{electrons available}: & \quad 3/4\ e^-\ (Ga^{db}) + 5/4\ e^-\ (As^{db}) = 2\ e^- \\ \text{electrons required}: & \quad 0\ e^-\ (Ga^{db}) + 2\ e^-\ (As^{db}) = 2\ e^- \end{aligned} \quad (2.1)$$

Although the vast majority of stable surface configurations fulfill the ECR, it is only an indicating condition, as stable configurations that do not fulfill the ECR have been found as well [64]. The determining factor is the minimization of surface free energy, which does not necessarily require a semiconducting surface.

2.3. The GaAs(0 0 1) surface

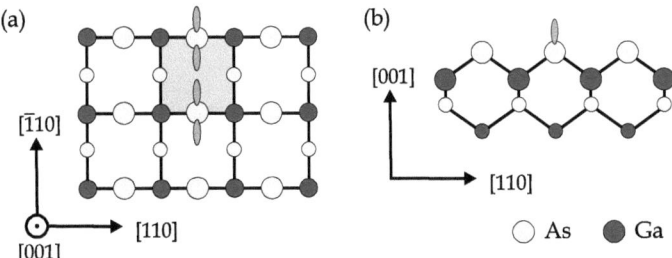

Figure 2.3: (a) *Top view and* (b) *side view of the GaAs(0 0 1) bulk-truncated surface. The surface unit cell is marked by a yellow square. The As dangling bonds are illustrated by gray ovals. Atoms below the figure plane are depicted by smaller circles. For reasons of clarity the lower Ga atom plane is not shown in the top view image.*

The bulk-truncated GaAs(0 0 1) surface, as shown in Fig. 2.3, is either group-III or group-V terminated. To lower the surface energy, rearrangement processes of the surface atoms lead to surface reconstructions that minimize the number of dangling bonds. Thereby the GaAs(0 0 1) surface shows a large variety of reconstructions, depending on the ambient circumstances, reaching from highly Ga-rich to extremely As-rich stages [65]. A schematic overview of the different reconstruction domains observed by RHEED is given in Fig. 2.4.

In principle two growth regimes for epitaxial growth of InAs on the GaAs(0 0 1) surface exist, separated by the phase transition of the GaAs(0 0 1) surface reconstruction from the As-rich GaAs-c(4×4) reconstructed surface to the As-rich GaAs(0 0 1)-$\beta 2(2\times 4)$ reconstructed surface between 490–510 °C [66]. However, this transition is not completely abrupt and thus hysteresis effects led to the assumption of an interstitial stage of (2×1) periodicity in the RHEED observations in Ref. [65]. Recently, the co-existence of both the c(4×4) and the $\beta 2(2\times 4)$ reconstruction was observed during the phase transition. These STM studies found no evidence for a stable interstitial reconstruction [67].

The c(4×4) and the $\beta 2(2\times 4)$ surface reconstructions are both of high relevance for applications and thus the basic substrate reconstructions for the studies in this work. Therefore both will be discussed here in more detail.

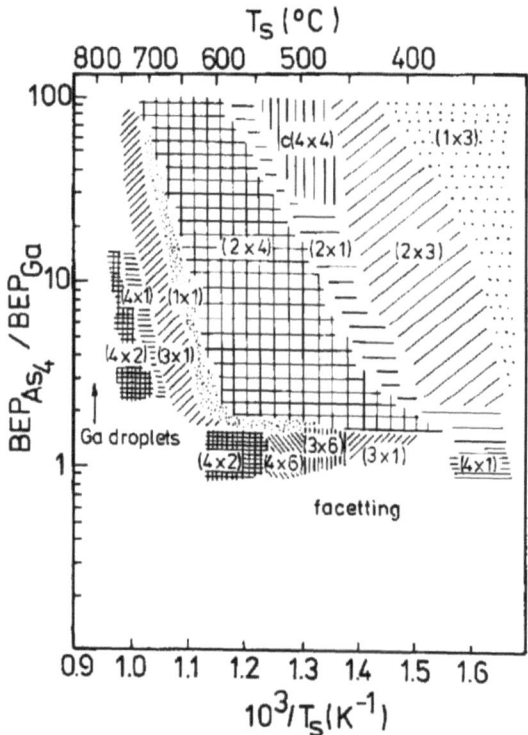

Figure 2.4: The surface phase diagram of reconstruction phases observed by RHEED on GaAs(0 0 1) in relation to the substrate temperature T_S and the As$_4$/Ga flux ratio, published in Ref. [65]. The narrow region of (2×1) structure between the c(4×4) and the (2×4) structure is ascribed to hysteresis effects of the transition.

2.3.1. The GaAs(0 0 1)-c(4×4) surface reconstruction

The larger part of the studies in this work is based on the GaAs(0 0 1)-c(4×4) reconstructed surface, which has become particularly important for the self-organized growth of InAs QDs [10, 68–72]. Forming under constant As-rich conditions in the rather low temperature regime of 480 °C or less [66], the c(4×4) reconstruction is the most As-rich phase of the GaAs(0 0 1) surface reconstructions. For reference studies especially at room temperature, this surface remains largely unchanged when prepared and quenched [73].

The commonly accepted structural model, as shown in Fig. 2.5 a, is generally described by three As dimers followed by one dimer vacancy along the [$\bar{1}$ 1 0] direction on top of the As-terminated bulk-truncated surface. The assumption of this structure was firstly based on grazing-incidence X-ray diffraction and STM images, the latter revealing the now well-known brick-like appearance of the triple dimer blocks [74–78]. Especially under extremely As-rich conditions, the c(4×4) reconstruction is characterized by a very low surface energy, comparable to natural cleavage surfaces like the GaAs(1 1 0) surface [57].

In order to determine if the As dimer model meets the ECR the electron balance of the available surface state electrons has to be considered. As illustrated in Fig. 2.5 a, the c(4×4)

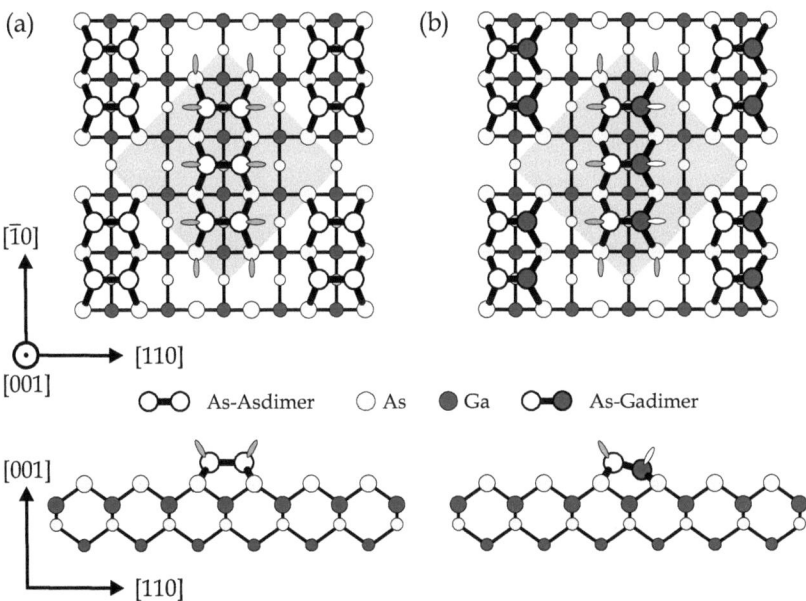

Figure 2.5: *Top views and side views of the GaAs-c(4×4) reconstructed surface, in (a) As–As dimer configuration and (b) Ga–As heterodimer configuration. Filled As dangling bonds are depicted by gray ovals and empty Ga dangling bonds are depicted by white ovals. Atoms below the figure plane are depicted by smaller circles. For reasons of clarity the lower Ga atom plane is not shown in the top view images.*

surface unit cell is characterized by a hollow site with four neighboring As surface atoms that each possess one unsaturated dangling bond and three As top dimers consisting of six As atoms in total that each possess one unsaturated dangling bond as well. Since each As dangling bond holds 5/4 electrons, there are $10 \cdot 5/4$ electrons available in total from the As dangling bonds. There are no Ga surface atoms and thus no Ga dangling bonds, which would hold 3/4 electrons each if present. Furthermore, the non bulk-like atomic bonds have to be considered as well. The six As atoms that form the three As top dimers are on top of an As-terminated bulk-truncated surface. Thus in addition to the total three As–As dimer bonds of the As top dimers, each single atom of the six As top atoms is bonded to the surface by two additional As–As back-bonds, summarizing to fifteen As–As-like bonds per surface unit cell. Since each atom of the As–As-like bonds contributes another 5/4 electrons, there are a further $15 \cdot 2 \cdot 5/4$ electrons available. According to the ECR, each As dangling bond and each As–As-like bond requires two electrons for saturation. As Eq. 2.2 shows, the situation of the surface state electrons in this structural model is balanced:

$$\begin{aligned} \text{electrons available:} \quad & 0 \cdot 3/4 \ e^- \ (Ga^{db}) + 10 \cdot 5/4 \ e^- \ (As^{db}) + 15 \cdot 2 \cdot 5/4 \ e^- \ (As\text{–}As) = 50 \ e^- \\ \text{electrons required:} \quad & 0 \cdot 0 \ e^- \ (Ga^{db}) + 10 \cdot 2 \ e^- \ (As^{db}) \quad\quad\quad + 15 \cdot 2 \ e^- \ (As\text{–}As) = 50 \ e^- \end{aligned} \quad (2.2)$$

Nevertheless an alternate surface model assuming Ga–As heterodimers (*c(4×4)-hd*) rather than As–As dimers was discussed as well (Fig. 2.5 b) [73,79,80]. It was shown that the ambient parameters during preparation have a significant influence on the final surface structure. The hd-structure was observed using As_4-molecules in the ambient atmosphere to stabilize the surface. When offering the more reactive As_2-molecules, the hd-structure was only observed as a metastable phase during the transition from the As-dimer dominated c(4×4) structure to the β2(2×4) reconstructed surface [66]. However, replacing one or two or all three As-dimers by As–Ga heterodimers in any case still fulfills the ECR. The final surface reconstruction type thus simply is a result of kinetic accessibility, and local interstages between the two models become very likely [56,81].

2.3.2. The GaAs(0 0 1)-β2(2×4) surface reconstruction

The second part of the studies in this work is based on the GaAs(0 0 1)-β2(2×4) surface, which is technologically important for the growth of sophisticated QD devices [12,82,83]. Under varying growth parameters (e.g. temperature and material fluxes) it remains widely unchanged in a stable configuration [65]. However, quenching the β2(2×4) reconstructed surface, e.g. for studies at room temperature, is rather challenging, as a saturated As atmosphere may easily cause the surface to transform into the c(4×4) configuration during quenching, and absent As might lead to more Ga-rich configurations, which cannot be reversed (cf. Fig. 2.4).

The As-rich GaAs(2×4)-reconstructed surface has been subject of many studies, yet the atomic structure remained uncertain for a long time. There have been several proposals of

2.3. THE GaAs(001) SURFACE

possible surface configurations, until the β2-configuration was finally confirmed by Ref. [84] using STM studies and first-principles calculations. Figure 2.6 shows the structural model of the GaAs-β2(2×4) reconstruction, as originally proposed by Ref. [85]. Two As dimers alternate with two dimer vacancies along the [110] direction. These vacancies lead to a trench along the [1̄10] direction, with a depth of one ML. In this trench another As dimer is formed, resulting in a configuration of three As dimers per unit cell. The β2-surface configuration meets the ECR (Eq. 2.3) and has been confirmed preferable over alternative (2×4) configurations for minimizing the surface energy [57, 86, 87].

$$\begin{aligned} \text{electrons available}: & \quad 4 \cdot \tfrac{3}{4}\,e^- \text{ (Ga}^{db}\text{)} + 6 \cdot \tfrac{5}{4}\,e^- \text{ (As}^{db}\text{)} + 3 \cdot 2 \cdot \tfrac{5}{4}\,e^- \text{ (As–As)} = 18\,e^- \\ \text{electrons required}: & \quad 4 \cdot 0\,e^- \text{ (Ga}^{db}\text{)} + 6 \cdot 2\,e^- \text{ (As}^{db}\text{)} \quad\;\; + 3 \cdot 2\,e^- \text{ (As–As)} = 18\,e^- \end{aligned} \quad (2.3)$$

It should be noted that many reports refer to temperatures for QD growth well within the range of the (4×4)-(2×4) phase transition [10, 25, 28, 30, 32, 88]. Yet, without detailed information on the measuring accuracy of the growth temperature or further structural analysis by e.g. RHEED, the actual surface reconstruction in these reports unfortunately remains unclear.

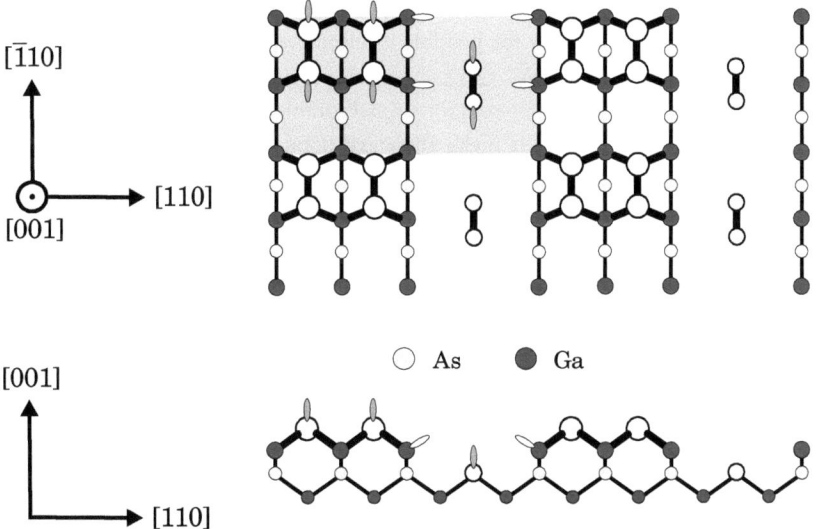

Figure 2.6: Top view and side view of the GaAs-β2(2×4) surface reconstruction. The surface unit cell is marked by a yellow rectangle, the filled As dangling bonds are depicted by gray ovals, the empty Ga dangling bonds are depicted by gray ovals. Atoms below the figure plane are depicted by smaller circles. For reasons of clarity the lower Ga atom plane is not shown in the top view images.

2.4. Semiconductor nanostructures

Low-dimensional semiconductor structures are typically created by embedding one semiconductor material into the matrix of another semiconductor material, usually of a larger band gap (*heterostructure*). If the geometrical extent of the embedded semiconductor is reduced in at least one spatial dimension below the order of the *de Broglie wavelength* of thermal electrons, the energy levels of the enclosed charge carriers no longer appear continuous, as the distance of the energy levels exceeds the thermal energy defined by $k_B T$. As $k_B T$ is hence insufficient for any thermal excitation of the charge carriers from the ground state to higher states, they are literally trapped within the embedded material [5, 8].

The de Broglie wavelength λ_B is defined in Eq. 2.4:

$$\lambda_B = \frac{h}{p} = \frac{h}{\sqrt{2m^* k_B T}} \, . \tag{2.4}$$

The determining parameters are the effective mass of the charge carrier m^* and apparently the ambient temperature T. For electrons in a typical III/V semiconductor like GaAs, λ_B yields 30 nm at room temperature, consequently LDS are of sizes less than some tens of a nanometer. Thus such structures are generally termed *nanostructures*.

The most important consequence of the reduction of spatial dimensionality (*confinement*) in nanostructures is the quantization of the electronic density of states $D(E)$, illustrated in Fig. 2.7, which is significant for possible electronic applications. Confinement in one dimension leads to *quantum wells* (QW), in two dimensions to *quantum wires* (QWR), and finally to *quantum dots* (QD) for three-dimensional confinement. Basically, size and shape of such nanostructures determine their electronic and optical properties [8].

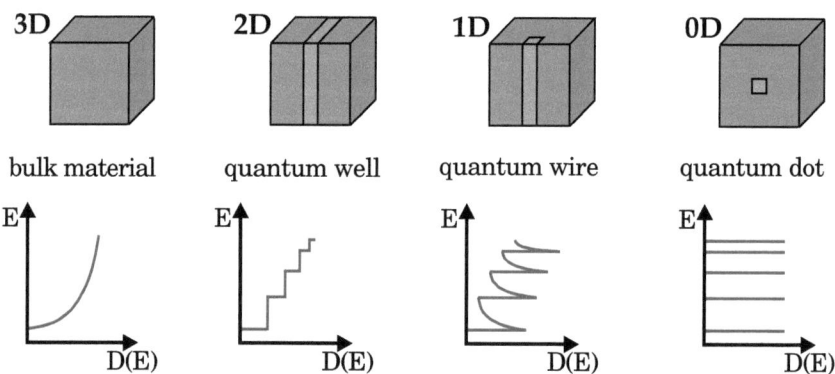

Figure 2.7: Confinement of charge carriers in heterostructures from three-dimensionality to zero-dimensionality and its consequences for the density of electronic states.

2.4.1. Epitaxial growth

Among several fabrication techniques for nanostructures, e.g. nanolithography [89] and etching techniques [90], self-assembling mechanisms in epitaxial crystal growth are widely used [8, 17, 91, 92]. The term *epitaxy* describes the formation process of a crystalline structure on an underlying crystal surface — the substrate — by depositing new material onto that substrate. The crystallographic orientation of the grown crystal is then determined by the substrate surface. If the deposited material is similar to the substrate material, this is commonly referred to as *homoepitaxy*. If the deposited material differs from the substrate material in e.g. crystal structure properties or chemical composition, this is commonly referred to as *heteroepitaxy*.

Basically, in heteroepitaxy near the thermodynamical equilibrium, three growth modes exist. The preference for a specific growth mode is determined by the total surface energy of the system. Therefore the energy of the grown material layer E_{lay} is considered in respect of the energy of the bare substrate surface E_{subs}. E_{lay} hereby includes the contributions of the surface energies of the newly formed surface E_{surf} and the interface between substrate and epilayer E_{int}, and the strain energy E_{str} caused by a possible lattice mismatch between substrate and epilayer. E_{str} increases directly with the increasing amount of deposited material up to a critical thickness, where the accumulated strain energy may cause lattice dislocations in the crystal structure.

In the case of $E_{lay} < E_{subs}$ the total coverage of the substrate surface by a film of deposited material is energetically favorable. This *Frank-van der Merwe* (FvdM) growth mode [93] (Fig. 2.8 a) is typical for materials with no or very little lattice mismatch.

In the opposite case, $E_{lay} > E_{subs}$, such film growth is energetically not favorable. Thus, three-dimensional islands grow on top of the substrate, by which the contributions of E_{int} and E_{str} can be reduced. This *Volmer-Weber* (VW) growth mode [94] (Fig. 2.8 b) is typical for materials with rather large lattice mismatch.

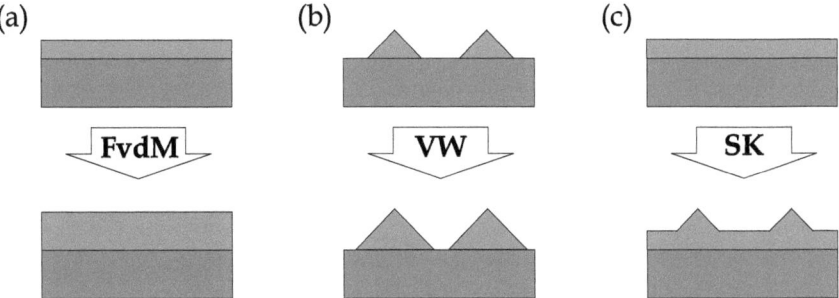

Figure 2.8: *Illustration of growth modes in self-assembled QD growth, (a) Frank-van der Merwe growth mode, (b) Volmer-Weber growth mode and (c) Stranski-Krastanow growth mode.*

In material systems with a moderate lattice mismatch, e.g. 7.2% for InAs on GaAs, a combination of both modes can occur. With the increasing thickness of a grown film — the so-called *wetting layer* (WL) — the accumulated strain energy E_{str} can result in an inversion of the energetic relation. As a result further material assembles in 3D islands evolving on top of that WL to reduce the elastic strain energy [37, 95]. This *Stranski-Krastanow* (SK) growth mode [96] (Fig. 2.8 c) is the typical growth method for the fabrication of QDs and has been investigated in detail for various material systems [35, 97–100].

2.4.2. Stranski-Krastanow growth of InAs/GaAs quantum dots

For the InAs/GaAs material system, it is well known that on the GaAs(0 0 1) surface QDs evolve in SK growth mode [36, 37]. The critical thickness for the 2D→3D transition from the WL to QDs is reported to be at about 1.6 monolayers (ML) of total deposited InAs material [35, 68, 101, 102]. This amount is rather independent of the actual growth conditions and the initial GaAs(0 0 1) surface reconstruction (the c(4×4) or the $\beta 2(2\times 4)$ configuration), on which the InAs typically is deposited. The lattice mismatch of the InAs WL to the GaAs(0 0 1) surface causes a significant accumulation of compressive strain energy with increasing thickness of the WL. This strain can partly relax in QDs upon the 2D→3D transition, lowering the strain energy E_{str} at the cost of increasing surface energy E_{surf}. Nevertheless, the total energy of the system minimizes.

The detailed structure of the resulting InAs/GaAs QDs has been in the focus of many investigations that revealed information about density, size, shape, and stoichiometry, both of free standing and overgrown QDs (e.g. [37–43]).

However, despite of the successes in QD characterization, the complex evolution of the first atomic layers of the deposited InAs on the GaAs(0 0 1) substrate is still not understood sufficiently [36]. Depending on the investigation techniques, different surface reconstructions for the initial WL have been reported. These range from (1×3) derived from *reflection high energy diffraction* (RHEED) patterns [71, 103] over (2×3) observed in *X-ray diffraction* (XRD) experiments [104, 105] to (4×3) based on *scanning tunneling microscopy* (STM), which unfortunately missed sufficient atomic resolution [106, 107]. Intermixing of deposited material with the underlying matrix has been assumed for low growth rates [108], while rather sharp interfaces are reported for faster growth rates, significant for device applications [109]. There is also a strong reduction of accumulated strain reported to occur at about 1.0 ML, leading to the assumption that this value may be the actual critical thickness [110].

3. Molecular beam epitaxy (MBE)

In crystal growth the most important precondition is a clean substrate with a stable surface configuration and little defect density to start with. Only this allows to meet the requirements of obtaining epitaxially grown thin material films, or epitaxially grown nanostructures in general, with their special electronic and optical properties. These properties can be easily affected yet by low crystal defect concentrations.

Metalorganic chemical vapor deposition (MOCVD) and *molecular beam epitaxy* (MBE) are the most common techniques for semiconductor epitaxy [111].

In MOCVD the source materials are metal organics and metal hydrides containing the chemical elements required for deposition. The substrate is mounted on a susceptor in a reactor chamber, and nitrogen or hydrogen are usually used as carrier gas for the material deposition. The epitaxial layer on the substrate surface is then formed by chemical reaction of the constituent chemicals [112–114].

In MBE, which is used in the present work, the deposition material is evaporated from elemental sources and is transported physically via a molecular beam onto the heated substrate surface.[1] The arrival rates of particles from the constituent elements hereby determine the chemical composition and doping level of the grown epilayer [117]. Today, MBE has emerged as a versatile technique for the growth of semiconductor nanostructures [118,119].

3.1. Ultra high vacuum (UHV)

It is evident from the beam nature of the material transport in MBE that interactions both among the molecular beams themselves and with the inevitable residual gas have to be minimized in order to avoid scattering processes and to preserve a stable and well-aligned material flux. These conditions will securely be ensured, if the mean free path of the beam particles exceeds the traversed distance between source and sample. In a standard MBE setup this distance can be estimated to about 0.2 m, so the simplest approximation specified for typical GaAs growth conditions yields $p_{rg,max} = 8 \cdot 10^{-4}$ mbar for the maximum value of the residual gas pressure [120]. *Ultra high vacuum* (UHV) is generally referred to as the pressure range of residual gas $p_{rg} \leq 10^{-9}$ mbar. Clearly, UHV is not necessarily needed to

---

[1] In addition to the most commonly used *solid source* MBE referred to here, other techniques of MBE exist, e.g. the *metalorganic molecular beam epitaxy* (MOMBE) [115] and *chemical beam epitaxy* (CBE) [116].

ensure a sufficient mean free path of the traversing molecules, but for epitaxial growth yet another condition has to be met.

Growth rates in MBE are comparatively slow, typically reaching from 0.01 ML/s to 1 ML/s. This ensures the impinging species on the surface to migrate to a favorable site in the growing crystal layer. However, at slow growth rates, residual gas particles may also penetrate the surface creating dopant effects, contaminations, and other growth defects. To avoid this, the concentration of residual gas particles near the sample surface must be negligible as compared with the concentration of beam particles; in other words, the time t_{rg} during which one ML of residual gas particles would be deposited on the surface must vastly exceed the time t_m of depositing one ML of growth material, e.g. $t_{rg} = 10^4 \, t_m$. Assuming typical growth conditions for GaAs in MBE with $t_m = 1\,\mathrm{s}$, then $t_{rg} \approx 2.8\,\mathrm{h}$ results [120].

The rate of impingement on a unit substrate area in unit time, e.g. j_{rg} for particles from the residual gas, is related to the partial pressure p_{rg} of the residual gas, the specific molecular weight M_{rg}, and the residual gas temperature T by Eq. 3.1 [120], with N_A and k_B being the *Avogadro* and *Boltzmann* constants, respectively.

$$j_{rg} = p_{rg} \sqrt{\frac{N_A}{2\pi M_{rg} k_B T}} \qquad (3.1)$$

Approximating the residual gas by nitrogen molecules, the condition for $t_{rg} \approx 2.8\,\mathrm{h}$ is met, when the residual gas pressure p_{rg} is as low as $1.7 \cdot 10^{-10}$ mbar [120], which demonstrates the importance of UHV conditions in MBE.

Slow growth rates and UHV condition in MBE allow the growth of structures with very low concentrations of crystal defects, smooth surfaces, and sharp material edges, as changing the material flux from the sources effects the sample growth instantly [121]. Furthermore, being a unique advantage to other epitaxial growth techniques, *in situ online* surface diagnostic methods, as e.g. *reflection high energy electron diffraction* (RHEED) (introduced in Sect. 3.5), can be used in MBE to analyze and control the growth process continuously.

3.2. Evaporation

The most essential requirement in epitaxial growth both for fundamental research and for applications is undoubtedly the uniformity and the reproducibility of growth results [118, 122, 123]. Repeating a growth process under exactly the same growth conditions must lead to the same results concerning e.g. stoichiometry, size, and shape of the epitaxial film or structure. Clearly, in MBE this reproducibility depends on the uniformity of the molecular beam. However, as discussed in Sect. 3.1, interactions between particles within the molecular beam are negligible because of their large mean free path lengths. Thus, the uniformity of the molecular beam primarily depends on the characteristics of the evaporation source.

In principle, an evaporation source in solid source MBE consists of a crucible filled with the desired material in the condensed phase in high purity. Heating the crucible then generates a temperature dependent molecular beam of material that can be interrupted by a shutter above the orifice to control the material flux. Eventually, the uniformity of the molecular beam depends on the isotropy of the material flux from the source, geometric parameters in the alignment between sources and sample, and the reproducibility of flux variations caused by temperature adjustments and shutter operations [124].

Evaporation sources that allow to control these conditions are so-called *effusion cells*. Their construction is based on a concept established by M. Knudsen resulting from the theory of evaporation [125]. Hence, especially in solid source MBE, the term *Knudsen cell* (K-cell) is most common.

3.2.1. Theory of evaporation

The first systematic investigations of evaporation rates in vacuum were conducted by *Hertz* in 1882 [126]. Following Ref. [120], *Hertz* investigated the evaporation losses of mercury when distilled in an evacuated enclosure and measured the pressure of the surrounding gas. *Hertz* concluded from his results that at a given temperature the ability of a surface to evaporate is limited to a specific maximum rate. Yet this maximum rate can only be achieved if the vapor pressure of the evaporating molecules compensates the equilibrium pressure p_{eq} on that surface without any of the molecules returning, which means that the hydrostatic pressure in the gas phase must be maintained to $p = 0$. Thus, the maximum number of molecules dN_e evaporating from a surface A_e during the time dt is derived from the rate of impingement j_{rg} at p_{eq} (Eq. 3.1) minus the contribution of returning molecules described by p (Eq. 3.2).

$$\frac{dN_e}{A_e dt} = (p_{eq} - p) \sqrt{\frac{N_A}{2\pi M k_B T}} \qquad (3.2)$$

However, *Hertz* did only manage to measure rates of about one tenth of the theoretical maximum rate. *Knudsen* later assumed that a fraction of vapor molecules $(1 - a_v)$ may be reflected at the gas–liquid interface of the evaporant surface [127]. Thus, these molecules contribute to the evaporant pressure, but not to the net material flux from the condensed phase into the vapor phase. *Knudsen* introduced the evaporation coefficient a_v to describe the ratio of the observed evaporation rate to the theoretical maximum rate given by Eq. 3.2. Generally referred to as the *Hertz-Knudsen* equation the more general form to describe the evaporation rate then is

$$\frac{dN_e}{A_e dt} = a_v (p_{eq} - p) \sqrt{\frac{N_A}{2\pi M k_B T}} . \qquad (3.3)$$

The Hertz-Knudsen equation also applies to evaporation from free solid surfaces, first shown by *Langmuir* investigating the evaporation of tungsten from filaments in evacuated glass bulbs [128].

3.2.2. Knudsen cells

Apparently, the maximum evaporation rate in Eq. 3.3 is yielded for $a_v = 1$ and $p = 0$. To obtain the maximum rate, *Knudsen* suggested the effusion from an isothermal enclosure with a small orifice, the Knudsen cell (Fig. 3.1). This enclosure — containing the evaporant material — would maintain the equilibrium pressure p_{eq} inside, if the orifice is very small compared to the evaporation surface. Furthermore, the diameter of the orifice must be very small compared to the mean free path of the gas molecules at p_{eq} inside the enclosure, and the thickness of its walls must be insignificantly small, so scattering or adsorption and desorption of molecules leaving the orifice is minimized. Then, the orifice might be regarded as an evaporating surface with the evaporant pressure p_{eq} and an evaporation coefficient of $a_v = 1$, corresponding to the inability of the orifice to reflect vapor molecules. The total effusion rate Γ_e, described as the number of molecules effusing from the orifice area A_e of the Knudsen cell, is then given by the Knudsen effusion equation

$$\Gamma_e \equiv \frac{dN_e}{dt} = A_e \left(p_{eq} - p\right) \sqrt{\frac{N_A}{2\pi M k_B T}} \tag{3.4}$$

and can be simplified by assuming $p = 0$ for effusion into UHV.

Nevertheless, in practice an effusion cell can only approximate the ideal Knudsen effusion. The walls of the cell orifice will always be of finite thickness, hindering the free molecular flow by a certain fraction and must be considered by further correction terms in Eq. 3.4.

Figure 3.1 shows the typical construction of a Knudsen cell. The evaporant material is filled into a crucible, which is often made of *pyrolytic boron nitride* (PBN). The crucible is enclosed by a heating stage and a watercooling shield to ensure the isothermality of the enclosure. The cell temperature T_c is monitored by a thermocouple. The diameter of the orifice is in the range of mm and an optional shutter can be used to interrupt the molecular beam. This standard type of a Knudsen cell is most suitable for the atomic evaporation of liquid metals such as gallium and indium, and has been used for the investigations in this work.

At standard evaporation conditions some elements like arsenic and phosphorus form tetrameric molecules (As_4 and P_4, respectively). These characteristics of the molecules evaporating in the molecular beam may have significant influence on the nature of the emerging sample surface [129]. As an example, As_4 or As_2 molecules impinging on the GaAs(0 0 1)-c(4×4) surface result in different atomic surface configurations [66]. Therefore the Knudsen cell for As evaporation used in this work has been additionally equipped with a heated cracker stage (Fig. 3.2), where As_4 molecules are disassembled into As_2 molecules by thermal decomposition. This process is temperature controlled. In the temperature zone above 900 °C at least 99.99% of the As_4 molecules are disassembled into As_2 molecules [129].

3.2. Evaporation

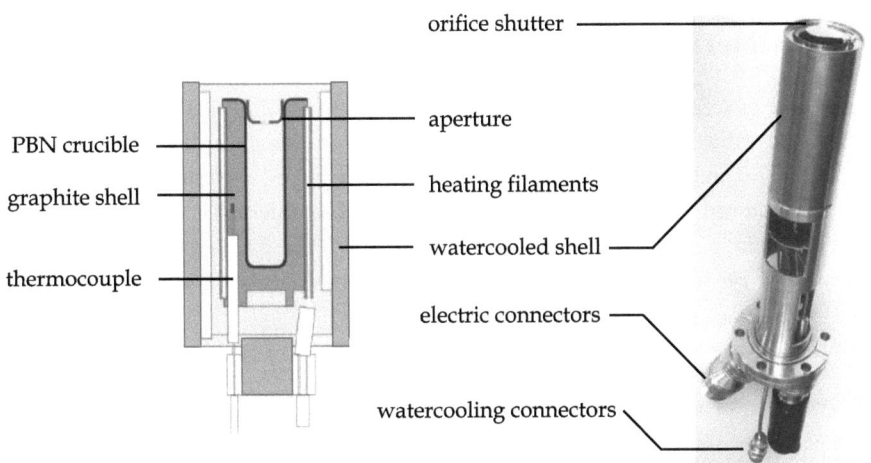

Figure 3.1: (left) Schematic layout of the evaporation stage in a common K-cell, (right) photograph of a commercial K-cell as used for group-III evaporation in this work.

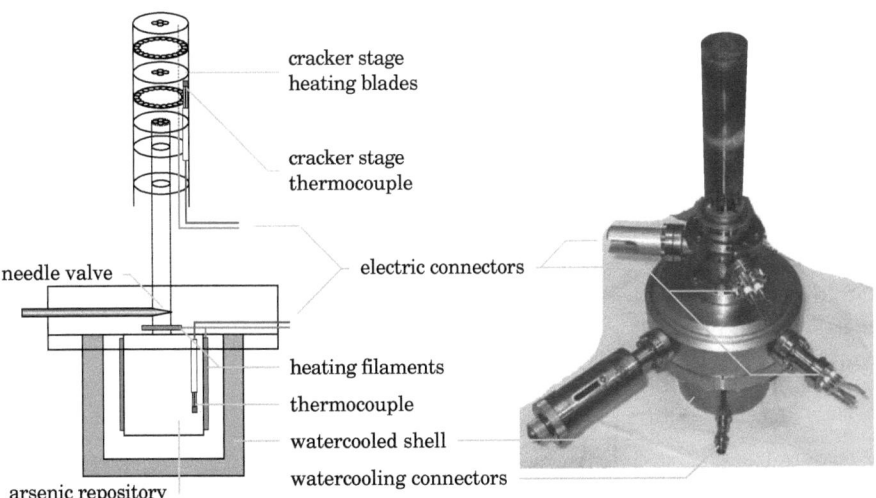

Figure 3.2: (left) Schematic layout of a cracker stage K-cell, (right) photograph of the cracker stage K-cell used for arsenic evaporation in this work.

3.3. Material distribution

3.3.1. Material flux

Atoms or molecules escaping the orifice will effuse into any direction within a hemisphere around the orifice. The particles enter the orifice from inside the cell with an angle ϑ and exit the orifice on the same trajectory of ϑ as scattering can be neglected at the orifice wall, which is assumed to be vanishingly thin (Fig. 3.3 a). The differential angular effusion rate from the orifice $d\Gamma_\vartheta$ is given by the *cosine law of emission* (Eq. 3.5, as derived from [120]). It is also known as the *cosine law of effusion* and equivalent to *Lambert's cosine law* in optics.

$$d\Gamma_\vartheta = \Gamma_e \cos\vartheta \, \frac{dS}{\pi r^2} = \Gamma_e \cos\vartheta \, \frac{d\Omega}{\pi} \qquad (3.5)$$

The molecular material flux to a unit surface area dS axially in front of the orifice in distance r is equal to the impingement rate j of particles on that surface area. This impingement rate can be defined as the differential angular effusion rate at $\vartheta = 0$:

$$j \equiv \left.\frac{d\Gamma_\vartheta}{dS}\right|_{\vartheta=0}. \qquad (3.6)$$

If the sample surface is mounted directly facing the orifice (Fig. 3.3 b), the angular material flux generally unfolds to:

$$j = \frac{\Gamma_e}{\pi r^2} \cos^2\vartheta \cdot \frac{r^2}{r'^2} = \frac{\Gamma_e}{\pi r^2} \cos^4\vartheta. \qquad (3.7)$$

Clearly, the angular dependence of the material flux induces an inhomogeneity of the material distribution towards the edge of the substrate area. Yet, for small substrates this is not significant, e.g. in the present setup the substrate holder in distance $r = 200\,\text{mm}$ has a diameter of 20 mm, which yields $\vartheta \approx 3°$ and thus $j(\vartheta) = 0.995\, j(0)$.

In practice, due to geometrical reasons, the effusion cells usually cannot be mounted directly facing the sample surface. When tilted by an angle φ (Fig. 3.3 c), the inhomogeneity of the material distribution increases, as the angular molecular flux further reduces by the cosine law to:

$$j_\varphi = \frac{\Gamma_e}{\pi r^2} \cos\vartheta \, \cos(\vartheta + \varphi) \cdot \frac{r^2}{r'^2}. \qquad (3.8)$$

To reduce the material inhomogeneity at the sample surface between the nearest point Ⓐ and the furthest point Ⓑ in Fig. 3.3 c, samples are usually rotated during film growth.

In the present work the MBE setup was equipped with three Knudsen effusion cells for the evaporation of gallium, indium, and arsenic. Spatial limitations opposite the sample holder required to place the effusion cells tilted by an angle $\varphi = 13°$. As the effusion cells are all tilted by the same angle φ, the magnitude of the additional reduction of the molecular

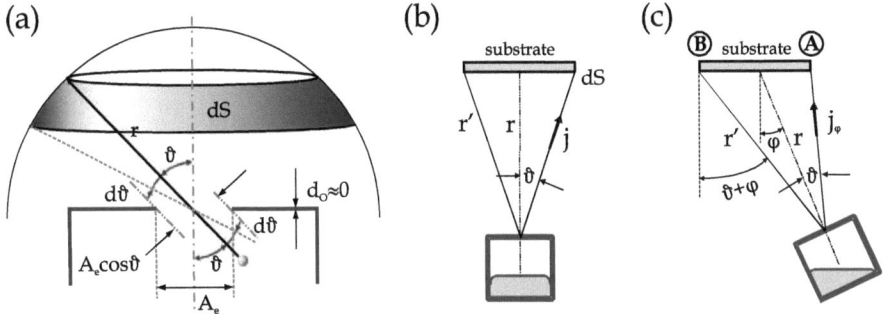

Figure 3.3: Illustration of the ideal evaporation from a Knudsen cell, (a) global material distribution from a perfect orifice; (b) ideal material flux onto a substrate surface from a source mounted axially to the sample or (c) mounted non axially, tilted by an angle φ.

flux j_φ can be regarded as equal for all effusion cells, which ensures the comparability of the beam parameters. An estimation of the variance of the angular material in this setup flux yields $j_{\varphi_A} = 1.006\, j_\varphi$ at point Ⓐ and $j_{\varphi_B} = 0.939\, j_\varphi$ at point Ⓑ, so that the rotation of the sample was considered unessential.

3.3.2. Beam equivalent pressure

In contrast to the ideal effusion cells described in theory, the description of real effusion cells often is hindered by the lack of detailed knowledge about diverging parameters to be considered in correction terms. These parameters may be the shape and the size of the orifice, the condensation of surplus material at the orifice, the fill level of the evaporant material, or the accuracy of the K-cell thermocouple [120]. In MBE practice the material flux from the effusion cells to the sample surface is therefore usually derived from the pressure of the molecular beam onto the sample surface, the *beam equivalent pressure* (BEP).

In the MBE setup for this work a movable ion gauge is placed at the sample position into the molecular beam to determine the BEP as a function of the cell temperature. The corresponding BEP data of the constituent effusion cells for arsenic, gallium, and indium can be found in Appendix A (Figs. A.1, A.2, A.3, respectively).

3.4. Surface growth mechanisms

As the impinging particles from the molecular beam arrive at the sample growth surface, various processes may take place (visualized in Fig. 3.4) [130]:

- physical adsorption of particles - *physisorption* (1)

- particle diffusion driven by kinetic activation - *surface diffusion* (2)

- conglomeration of particles - *nucleation* (3)

- incorporation of particles into the crystal structure - *chemisorption* (4)

- (thermal) *desorption* (5)

The significance of each process strongly depends on the ambient growth parameters. It is evident that for epitaxial growth the chemisorption must overbalance any desorption processes, so molecular beam and sample surface must be slightly off the thermodynamic equilibrium and a steady surplus material flux to the sample must be guaranteed. Therefore, the sample temperature T_s is usually lower than the effusion cell temperature T_c.

The probability of arriving particles to remain on the surface is described by the sticking coefficient α defined as the ratio of adsorbed particles N_{ad} to the total number of arriving particles N_i.

$$\alpha = \frac{N_{ad}}{N_i} \qquad (3.9)$$

Generally, if not desorbed instantly, impinging particles are tied to the surface by *van der Waals interaction* first. This physisorption does not involve any electron transfer between the adsorbed atoms or molecules (*precursors*) and the surface. Furthermore, its probability is less influenced by the local stoichiometry at the arrival site. The distance between adsorbates and surface is comparatively large to the atomic distances within the crystal structure, thus the linkage is not very tight and the adsorbates are rather free to move on the surface.

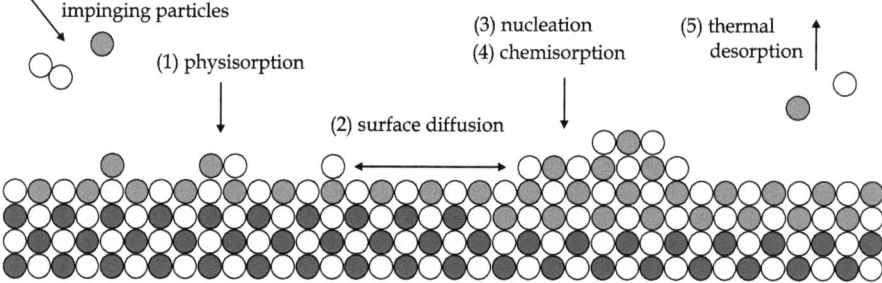

Figure 3.4: Illustration of principle surface growth mechanisms in MBE.

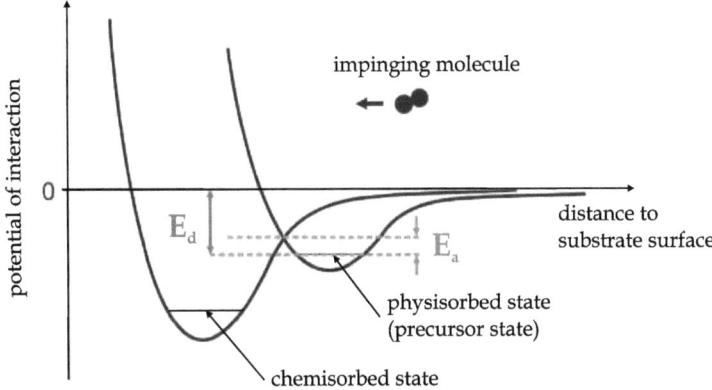

Figure 3.5: Potential of interaction between an impinging molecule and the adsorbing surface, following Ref. [131].

Driven by kinetic activation, the adsorbates diffuse on the surface to find a suitable site for chemisorption, i.e. the chemical reaction between adsorbate and adsorbent involving electron transfer. The probability of the chemisorption process depends highly on the local stoichiometry and configuration at the surface [132]. Furthermore, it depends on the chemical configuration of the precursors, i.e. being single atomic or molecular. Molecular precursors may require atomic dissociation before chemisorption. Chemisorbed incorporation and thereby crystal growth often starts at the edges of surface steps or nucleating islands, as more next neighbors are available at the crystal sites [132, 133]. As denoted in Fig. 3.5, the transition from the physisorbed state to the chemisorbed state may require some activation energy E_a, which should be less than the energy required for desorption E_d. If this is not the case and $E_a > E_d$, this rather leads to the desorption from the precursor state. The sample growth temperature T_s hereby is the adjustable parameter, and usually only a small window for temperature adjustments exists to ensure crystal growth.

3.5. Reflection high energy electron diffraction (RHEED)

The UHV setup of solid source MBE allows the application of *in-situ* surface diagnostic methods like *reflection high energy electron diffraction* (RHEED) [134, 135]. RHEED is a powerful tool to determine the surface reconstruction and structure anytime before, during, and after the growth process, as well as to monitor the growth process itself *online* [136–149].

The principal setup is shown in Figure 3.6. An electron beam is directed onto the sample surface at a shallow angle ($\approx 3°$) and reflected to a screen on the opposite site creating a diffraction pattern there. Due to the shallow incident angle, the penetration depth of the electron beam is very low, making RHEED highly surface sensitive [135, 140, 150]. The energy of the electrons in the beam is in the range of 10 keV corresponding to a *de Broglie* wavelength of about 0.012 nm.

3.5.1. Formation of diffraction patterns

The formation of the diffraction patterns can be ascribed to the elastic theory of scattering and may be explained by the use of a geometrical construction, the *Ewald sphere* [120, 151]. The arriving electron beam is elastically scattered at the crystal surface by atoms of a periodic lattice. The scattering amplitude now is at maximum, if constructive interference occurs. The criterion for constructive interference is given, if the scattering vector $\Delta\vec{k}_s$, i.e. the change between incoming wave vector $\vec{k}$ and outgoing wave vector $\vec{k}'$, equals a lattice vector of the reciprocal lattice $\vec{G}$ (*Laue condition*, Fig. 3.7):

$$\Delta\vec{k}_s = \vec{k}' - \vec{k} = \vec{G} \,. \tag{3.10}$$

As the law of energy conservation applies, the total amounts of the incoming $\vec{k}$ and the outgoing $\vec{k}'$ remain equal. If a sphere is defined in k-space with radius k to visualize all possible directions of $\vec{k}'$ it becomes evident that constructive interference occurs anytime a lattice point of the reciprocal lattice lies on the surface of this sphere (*Ewald construction*, Fig. 3.7).

Due to the two-dimensionality of a surface, its reciprocal lattice exhibits rods in z-direction. Hence, the intercept points of these rods with the surface of the Ewald sphere form semicircles in screen projection (*Laue circles*, Fig. 3.8). In practice, however, the observed RHEED patterns of a surface reconstruction rather consist of smeared lines than of sharp spots.

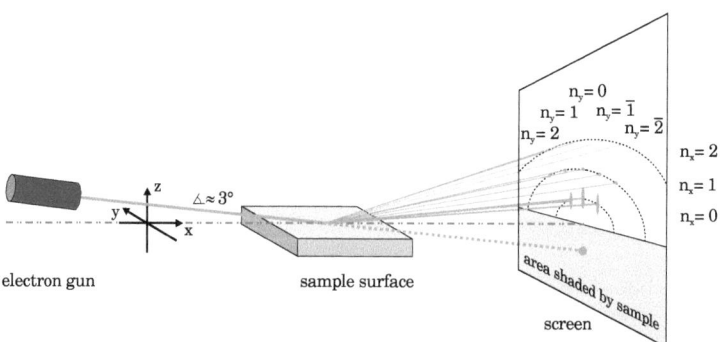

Figure 3.6: Schematic illustration of the RHEED principle. An electron beam is directed in shallow angle onto the sample surface, the reflected beam shows diffraction patterns of order $n_{x,y}$ forming characteristic Laue circles.

3.5. Reflection high energy electron diffraction (RHEED)

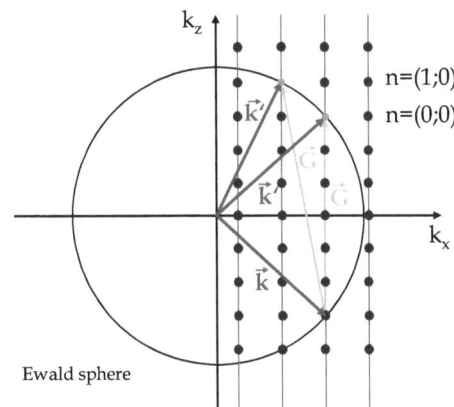

Figure 3.7: *Principal Ewald sphere construction and illustration of the Laue condition in RHEED.*

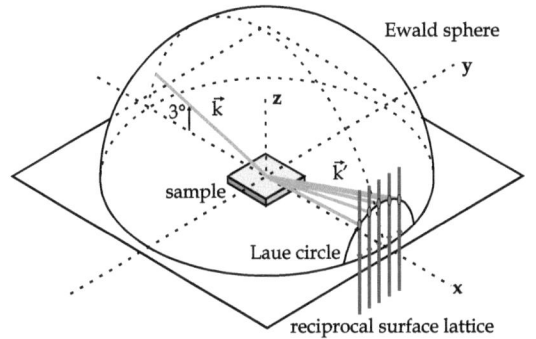

Figure 3.8: *Formation of Laue circles in RHEED.*

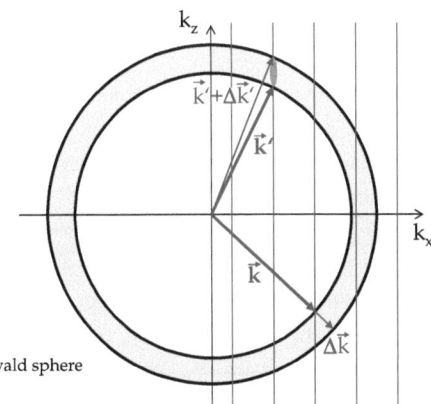

Figure 3.9: *The energy uncertainty causes a broadening of diffraction patterns.*

This must be ascribed to the uncertainty principle applying to the energy and angle of the electron beam. The inevitable energy distribution effectuates a distribution of the radius of the Ewald sphere, causing the reflexes in the diffraction pattern to extend in particular along the z-direction (Fig. 3.9). The angular distribution causes a similar broadening of the reflexes, yet it is much less significant.

The shallow incident angle, which allows RHEED to observe the sample surface during growth without blocking the molecular beam, also has a disadvantage. Patterns of adequate intensity can only be observed for certain surface directions, as the information of the perpendicular direction is covered in Laue rings of higher order with much less intensity. Thus, in order to obtain the full surface information, the sample surface is usually rotated.

RHEED is sensitive to the global surface ordering, as the diffraction information is averaged over the area covered by the electron beam. It is not possible to reliably detect minor local variations at the surface or even single objects. Yet, if these objects or variations are distributed across the covered surface area with sufficient statistical relevance, additional features may be observed in the diffraction patterns. Surface steps would interrupt the periodicity and cause a splitting of diffraction spots. Three-dimensional islands create an overlaid 3D superlattice, that causes overlaid single spots in the diffraction patterns. Furthermore, if QDs with a certain facet configuration evolve from these islands, wedge-shaped reflexes may appear (so-called *chevrons*). The apex angle of such a chevron allows to derive the orientation of these facets [152, 153].

3.5.2. Growth control

Its capability to monitor surface changes even during the growth process makes RHEED very suitable to control the growth process itself [154]. The quality of the growth surface hereby determines the intensity of the diffraction reflexes. Monitoring the specular diffraction reflex ($n_{x,y} = 0$) while growing at a constant growth rate, periodic oscillations of the reflex intensity become observable [137–139, 142]. The cause of this behavior is illustrated in Fig. 3.10, following [155–157].

A perfectly plain surface induces perfect constructive interference leading to a maximum intensity of the diffraction reflex (1). As surface growth starts, nucleation sites form, inducing island growth and the ordering of the surface reduces. This disorder scatters the diffraction beam and broadens the reflex, thus the reflex intensity decreases (2). In addition, destructive interference between the beams reflected at the surface of the substrate and the nucleated islands may further reduce the intensity. The maximum disorder on the surface and therewith the minimum intensity corresponds to half a monolayer of deposited material (3). Further deposited material starts to fully cover the original surface, reducing the scattering sites (4) until the complete new monolayer has evolved (5). With continuing growth this process proceeds periodically, allowing to withdraw information about the growth rate from the time interval between the intensity maxima.

3.5. REFLECTION HIGH ENERGY ELECTRON DIFFRACTION (RHEED)

Due to the statistical nature of the growth process, the maximum intensity decreases with time. The initial perfect flatness of the surface cannot be restored during growth, as variations in the particle allocations induce an increasing surface roughness. The incorporation of growth defects remaining within the crystal structure may also contribute to this effect. To restore a flat surface, the growth process has to be interrupted and the surface must be given time and kinetic ability (*annealing*). This annealing process can be monitored by RHEED as well.

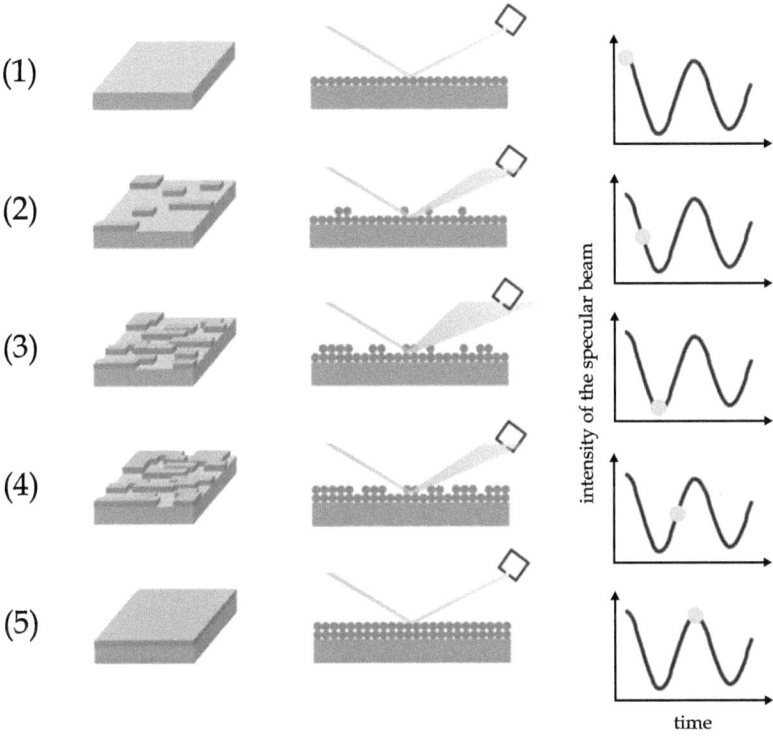

Figure 3.10: Schematic illustration of the origin of RHEED oscillations, following [155–157].

4. Scanning tunneling microscopy (STM)

In 1982 Binnig, Rohrer, Gerber, and Weibel developed *scanning tunneling microscopy* (STM) [1] as a revolutionary new technique for the investigation of crystal surfaces [158,159]. The novelty in this technique was the ability to image surfaces with actual atomic resolution directly in real space. Furthermore, from the image information the characteristics of the local surface structure, defects, adsorbates, and even the local chemical composition can be derived. To honor their invention Binnig and Rohrer were awarded the *Nobel Prize in Physics* in 1986.

Today, STM has become a well established technique in surface characterization and essential to unravel the details of structures used in modern nanotechnology. Involved in a steady progress of development it has been adapted, e.g., to investigate buried structures (XSTM) [2] or local electronic properties (STS) [3]. Based on STM a wide field of related scanning probe techniques has been developed, most notably AFM [4] and SNOM [5].

4.1. The STM principle

The STM technique basically utilizes the quantum tunneling effect. A sharp metallic tip is placed above the surface of a conducting or semiconducting sample with a small vacuum gap of distance s in between. There is a finite probability for electrons to tunnel through this vacuum barrier, which is determined by the size of the overlap of the electron wave functions of tip and sample. By reducing the vacuum barrier to the range of only a few *Ångstrøm* 1 Å = 0.1 nm this overlap becomes significant, allowing electrons to tunnel through the barrier from the tip to the sample or vice versa (Fig. 4.1). If there is an additional voltage V_T applied between sample and tip, a finite tunneling current I_T can be measured. Following Ref. [165], then

$$I_T \propto e^{-2\kappa s} \quad \text{with } \kappa = \sqrt{2m_e \Phi \hbar^{-2}}, \tag{4.1}$$

---

[1] STM is also used as abbreviation for *scanning tunneling microscope*

[2] *cross-sectional scanning tunneling microscopy / microscope* [160]

[3] *scanning tunneling spectroscopy* [161,162]

[4] *atomic force microscopy / microscope* [163]

[5] *scanning nearfield optical microscopy / microscope* [164]

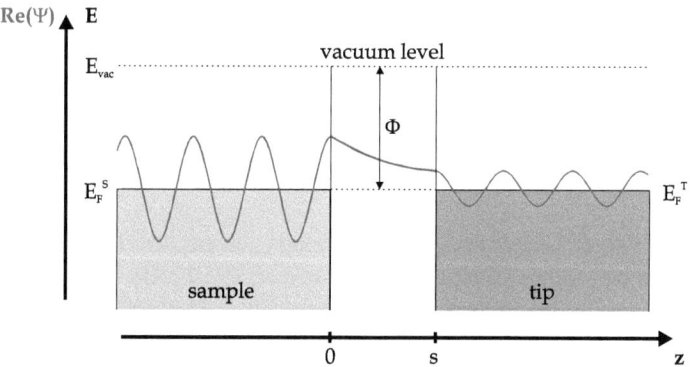

Figure 4.1: Basic model of one-dimensional tunneling in STM.

corresponds to the simplest approximation for a rectangular potential barrier of constant height Φ and results in a constant factor κ.

Assuming a barrier equal to the typical work function for semiconductors $\Phi \approx 4\,\text{eV}$ yields $\kappa \approx 1\,\text{Å}^{-1}$. Thus I_T is highly dependent on the distance s between tip and sample, varying by about one order of magnitude for a distance change of $\Delta s = 1\,\text{Å}$. Depending on the conditions of the sample surface, spatial resolution of $0.1\,\text{Å}$ vertically and $1\,\text{Å}$ horizontally can be achieved.

4.2. Surface imaging

In order to image a surface area, the STM tip is scanned line by line across the sample surface while the corresponding tunneling current is measured. The common principle is sketched in Fig. 4.2. The very fine movements of the tip, required for high spatial resolution, is realized by the use of piezo crystals, and usually an electronic feedback circuit is used to control the distance s between surface and tip by keeping the tunneling current I_T constant. Instead of the piezo tripod, as in the original setup in Fig. 4.2, a piezo tube is used in the setup of this work. A computer then generates a topographic image of the respective surface positions (x, y) and the corresponding values of the tip height variations Δs in scales of a brightness contrast.

The common theoretical approach to describe the tunneling process is the *Tersoff Hamann* approximation of *Bardeen's* theory of tunneling [166]. Herein an ideal tip is assumed with a single atom in front that only contributes s-orbital shaped electron wave functions and a constant density of states ρ_{tip}. Then the tunneling current I_T is approximately determined by the local density of states of the sample $\rho_{s,loc}$ at the tip location $\vec{r}$ integrated from the

4.2. SURFACE IMAGING

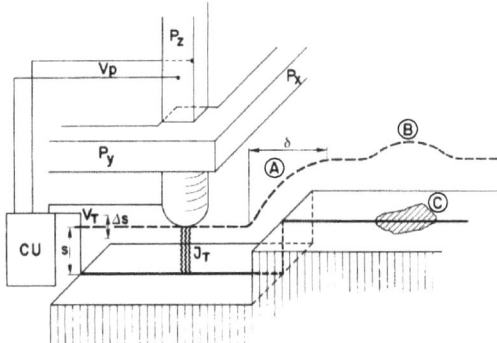

Figure 4.2: Operation principle of the STM, as originally developed by Ref. [158]. The metallic tip is scanned over the surface in distance s using a piezo tripod P_x, P_y and P_z. A control unit (CU) provides the tunneling voltage V_T and controls the tunneling current I_T to be constant by an appropriate piezo voltage V_p to adjust the tip height s concurrently. The variation of the tip height Δs then represents the actual STM contrast signal (dashed line).

Fermi level energy E_F to $E_F + eV_T$,

$$I_T \propto \int_{E_F}^{E_F+eV_T} \rho_{s,loc}(\vec{r},E)\, dE. \tag{4.2}$$

Thus, the STM signal is actually a measure of $\rho_{s,loc}$, the local density of states at the sample surface. More detailed descriptions on the theory and application of the STM principle are given in various textbooks, e.g. [167,168].

4.2.1. Modes of operation

The STM can be operated in different modes, depending on the experimental needs.

In **constant height mode**, the STM tip is scanned over the surface in constant distance s and the variations of the tunneling current ΔI_T are measured and directly converted into a surface image (Fig. 4.3 b). This mode allows very fast and sensitive surface scanning, but is only applicable on very flat surfaces. As the height distance s is a fixed value, the tip will be in risk to collide with any surface objects exceeding this height.

All the investigations in this work were obtained in **constant current mode**, which is the most commonly used mode in STM. Hereby the tunneling current I_T is set at a given value (typically 200–500 pA). A feedback loop then permanently adjusts the tip distance s while scanning according to the local surface characteristics, so that I_T remains constant (Fig. 4.3 a). In the setup of this work the height regulation is carried out by controlling the voltage of the z-component of the tube piezo. The information from the height adjustments Δs is then interpreted into a surface image.

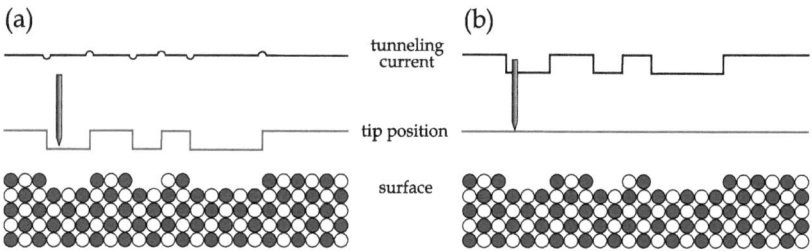

Figure 4.3: *Schematic principles of the operational modes in STM, (a) constant current mode (b) constant height mode.*

The advantages of the constant current mode are obvious. The constant readjustment of the tip height allows fast and stable surface scanning and avoids collisions with protruding surface objects. Still, the feedback loop takes some time to respond to surface changes. So any object is displayed with some delay and sudden changes like sharp step edges may be imaged a little blurred.

4.2.2. Contrast mechanisms

In principle, the visualized brightness contrast in the resulting STM images is a combination of two different types of contrast effects [54, 167].

The **topographical contrast** refers to the actual height profile of the sample. It is clear from Eq. 4.1 that a lowered distance s implies an increase of I_T, thus higher objects on the surface create brighter contrast in the STM image. The high sensitivity of the measured current I_T allows to observe terraces with monoatomic steps, atomic mesas and trenches, and even single atoms.

However, as the tip can never be of infinite sharpness, the observed data always exhibit a convolution with the tip geometry (Fig. 4.2 Ⓐ). At steep terrace steps, for instance, tunneling from side atoms of the tip cone near the edge additionally contributes to the tunneling current, thus the resulting contrast often rather displays the shape of the tip than the shape of the step edge.

Another aspect of the topographical contrast is due to strain in the crystal structure and plays an important role in particular in XSTM studies. Embedded material, like InAs QDs in a GaAs matrix, suffers from high compressive strain due to lattice mismatch. This strain cannot be relieved efficiently, unless situated near a surface, e.g. caused by a cleavage plane. In this case, the strained material may relax by bending outwards the cleavage surface by several Ångstrøm.

The other type of contrast effect that has to be considered when interpreting STM images is the **chemical and electronic contrast**. One chemical contrast variation to the tunneling current I_T is caused by the material specific work function W affecting the barrier height Φ

in Eq. 4.1. As illustrated in Fig. 4.2, a variation in the STM signal Ⓑ is observed at the site of a contamination spot Ⓒ of material with a lower W.

There is another contribution to the contrast due to the local electronic states of the sample, which directly affects the value of the tunneling current I_T at a given sample potential V_T, as stated in Eq. 4.2. Basically there are two possible scenarios for V_T. For positive sample bias $V_T > 0$ electrons would tunnel from the occupied states of the tip to non-occupied states of the sample. Imaging the surface of the III/V compound semiconductor GaAs, the STM signal then is Ga sensitive (*empty state image*). For negative sample bias $V_T < 0$ electrons would tunnel from the occupied states of the sample to non-occupied states of the tip, and the STM signal becomes As sensitive (*filled state image*).

Yet all images of the GaAs(0 0 1) surface presented in this work are filled state images, as there unfortunately was no success in obtaining empty state images. This might be due to the fact that the observed surface reconstructions were all highly As terminated (cf. Sect. 2.3). This probably impedes tunneling electrons from effectively tunneling into the empty states of the underlying Ga atoms.

For non-ideal tips there is always a certain roughness at the tip cone, and several front atoms may form so-called *microtips*, which all contribute to the tunneling current. Observed objects then might be imaged twice, when the tunneling current switches from one micro tip to the other. This process is referred to as *double* or *multiple tip effect*.

5. Experimental setup

5.1. The UHV chamber system setup

In this work a combined UHV chamber system was used to perform all investigations *in situ*, i.e. without exposing the samples to non-UHV conditions. The UHV chamber system was kindly provided by *Karl Jacobi* from *FHI Berlin*[1], under whose supervision it originally was developed [169]. The setup was especially designed for the surface investigation of MBE grown InAs/GaAs semiconductor samples, e.g. investigations of high-index stable GaAs surfaces [58,60,170,171] and detailed investigations of InAs/GaAs QDs [39,40,172,173].

Figure 5.1 shows a photograph and a simple sketch of the UHV chamber system, which basically consists of four modules separated by UHV valves: the MBE chamber, the sample storage, the STM chamber, and a chamber for preparation and further analysis. The horizontal transfer of samples between these stages is ensured by a long-stroke magnetically-coupled linear positioner (*transfer rod*), able to mount the sample carrier.

For maintenance the UHV chambers can be vented separately via external valves using dry N_2 gas, which reduces contaminations by air moisture and dust particle during this procedure. To restore the UHV conditions after venting each chamber is equipped with a system of combined pumping stages, typically consisting of a membrane pump/rotary vane pump, a turbo molecular pump, and a sputter ion pump. The respective chamber is baked at a temperature above 120°C for 1–3 days to support the removal of the residual gas and especially air moisture, that have entered and adhered to the chamber walls while being exposed to ambient air.

5.1.1. The MBE setup

The MBE chamber is equipped with three water-cooled solid-source Knudsen effusion cells for the evaporation of gallium, indium, and arsenic. A schematic sketch is given in Fig. 5.2. The effusion cells are mounted at the lower side of the chamber facing upwards to the sample carrier position. This setup is necessary, as many solid materials like Ga and In do not sublime but evaporate from the liquid phase. Obviously, at the optimum central position there can be only one effusion cell directly facing the sample, which would require the other two effusion cells to be placed with a significant offset angle φ. Such

---

[1] Fritz-Haber-Institut der Max-Planck-Gesellschaft

52   5. Experimental setup

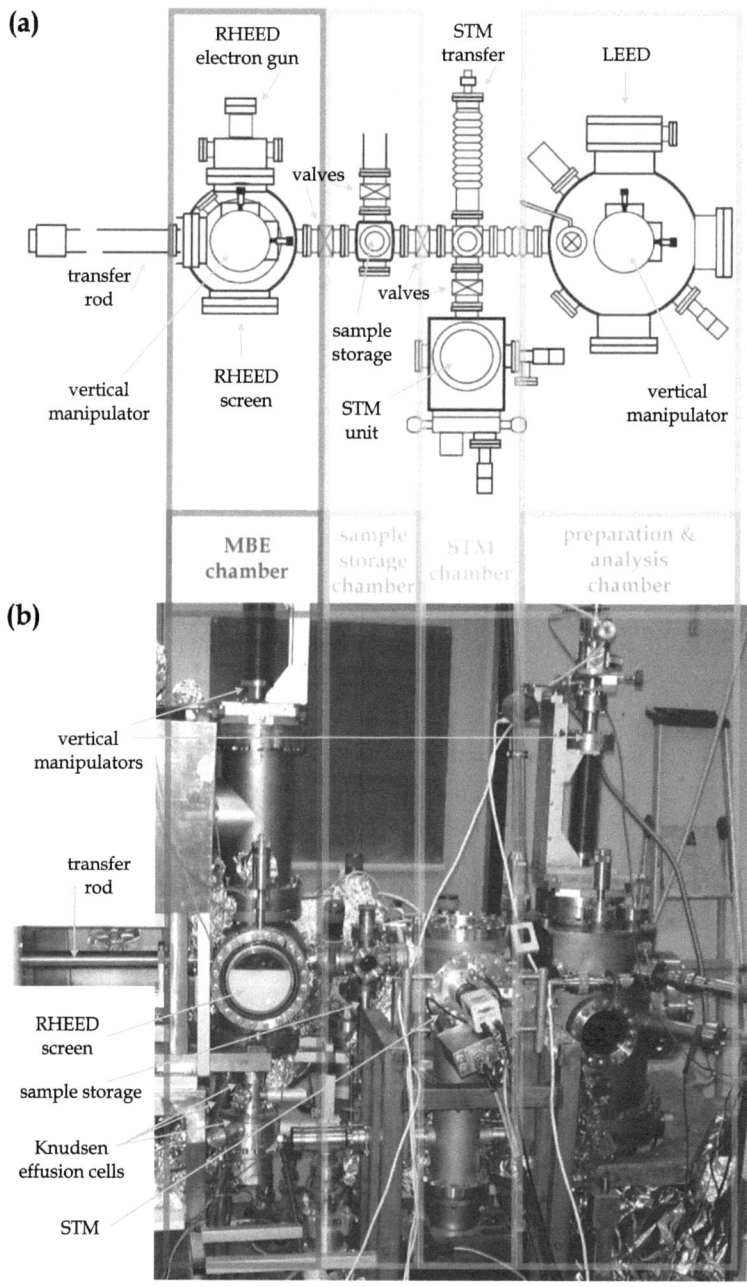

Figure 5.1: *(a) schematic top view and (b) photographic side view of the experimental UHV setup used in this work [169].*

an arrangement would lead to rather high inhomogeneities of the material distribution for these two sources, as discussed in Sect. 3.3.1. To reduce such effects, all sources were placed symmetrically around the center position with a common tilt angle of $\varphi = 13°$. The additional position for a possible fourth K-cell is used for the sample temperature control by an optical pyrometer at the moment.

For MBE growth the sample carrier is mounted onto a vertical manipulator with the sample facing downwards to the orifices of the effusion cells. The manipulator in particular allows remote rotation and vertical transport of the sample carrier. A filament is placed within a pocket on the backside of the sample carrier for radiation heating. The substrate temperature T_S is controlled by the value of the filament current I_F (see Appendix C for respective data).

In order to obtain the actual material flux arriving at the sample, a movable ion gauge can be placed exactly at the sample position and measure the BEP. In horizontal alignment to the sample there is the electron gun and the detection screen window for the RHEED setup. This window is partly uncovered to serve as well as an observation window when transferring samples. A mechanical shutter is used during growth to protect screen and window from the reactive molecular beam material from the inside.

This central part of the MBE chamber is enclosed within a cryopanel for cooling by liquid nitrogen (ℓ-N_2). In the upper part a sputter ion pump and an additional ion gauge are used to maintain and control the UHV conditions. Typical values for the residual gas pressure are in the $p_{rg} = 10^{-10}$ mbar range while using ℓ-N_2 cooling.

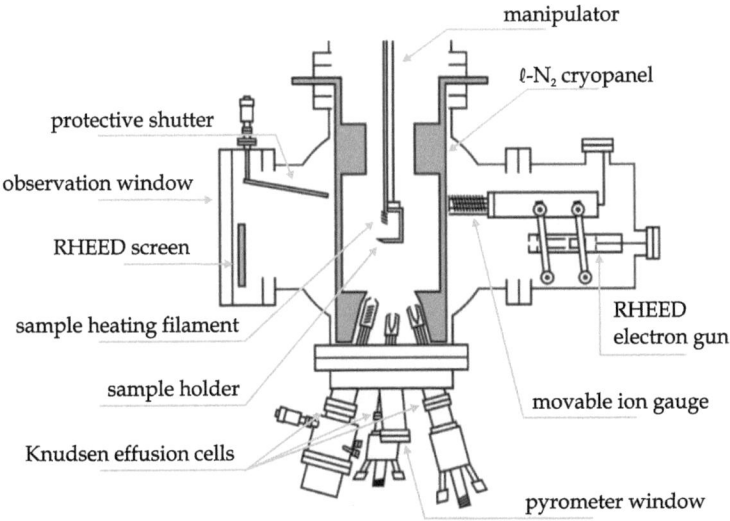

Figure 5.2: Schematic sectional side view of the MBE chamber setup, [169].

5.1.2. The STM setup

The STM itself is a commercial setup from *Park Scientific Instruments*, designed for horizontal tip–sample approach. A schematic sketch of the STM setup is given in Fig. 5.3.

The rather stiff and compact layout supported by a system of soft springs and eddy current damping decouples the setup from external vibrations. There is a mechanism to lock the STM, when sample or tip have to be replaced. The sample carrier is mounted horizontally onto a holder profile by a linear positioner. Inside the STM chamber there is a storage that can hold up to five tips. An additional filament and high-voltage connection allow tip cleaning by electron bombardment heating. A rotatable grabber can hold the tips and move them to the required position.

For the tip–sample coarse approach a walker unit is used. Three stacks of cross-paired shear piezo crystals hereby serve as a motor in slick-and-slide technique. The coarse ap-

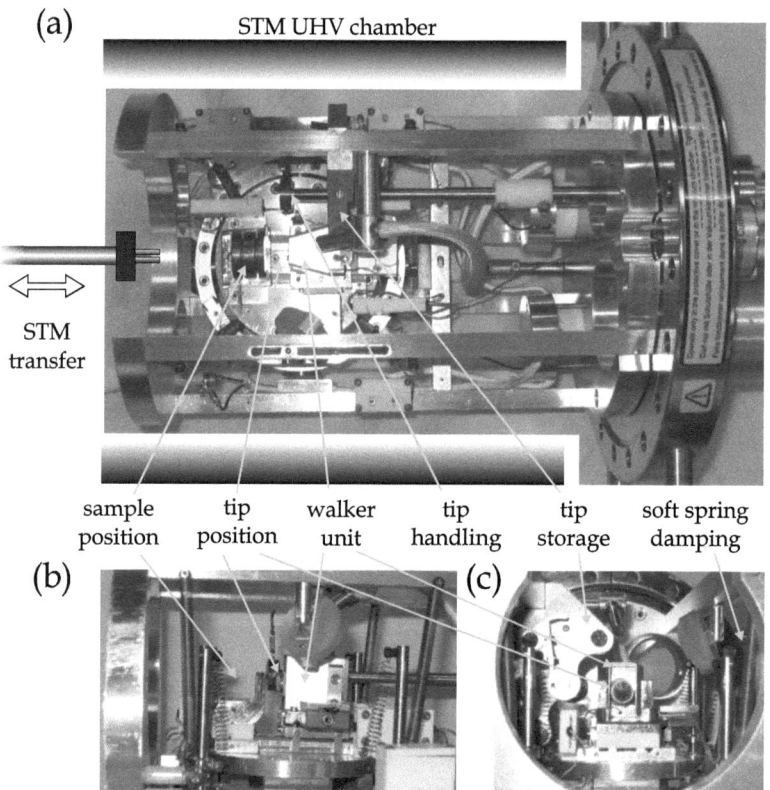

Figure 5.3: Photographic (a) top view, (b) side view, and (c) front view of the STM setup.

proach can be observed by a camera focusing on the sample surface. A sectorized tube piezo is attached in front of the walker for the fine positioning and scanning of the tip. The tip is mounted magnetically in front of that tube piezo.

The UHV conditions are maintained by a sputter ion pump attached to the bottom side of the chamber. Typical values for the residual gas pressure in the STM chamber are less than $5 \cdot 10^{-11}$ mbar.

5.1.3. The setup for preparation and further analysis

The UHV chamber setup for sample preparation is sketched in Fig. 5.4.

This chamber is equipped with another vertical manipulator to position the sample carrier in front of an ion gun. Here, the sample surface can be prepared by argon ion sputter etching. A needle valve allows the controlled introduction of Ar gas into the chamber. Ar atoms are then ionized by electron bombardment and accelerated by an electric field to traverse to the sample surface. The impinging Ar^+ ions thereby remove surface material, including surface oxides and adsorbates.

The UHV conditions are upheld and restored by the use of a turbo-molecular pump. Typical values for the residual gas pressure are around $1 \cdot 10^{-10}$ mbar.

In this setup there is an option for LEED[1] analysis, yet it was not used in the present experiments. Additionally there is a rotatable electron energy analyzer, which has been used previously for ARPES[2] experiments.

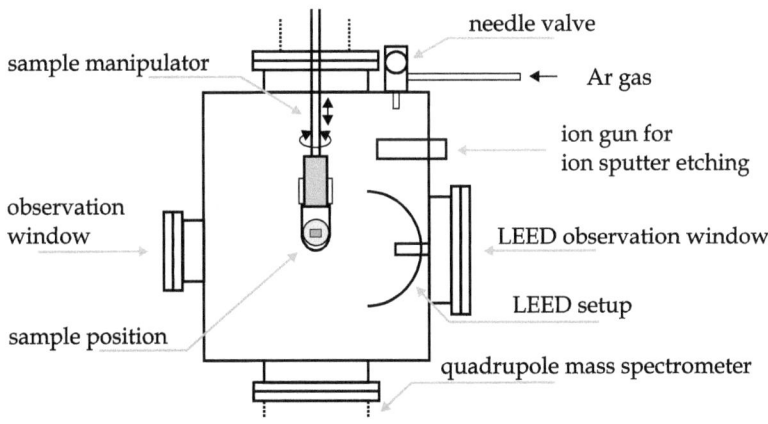

Figure 5.4: Schematic sectional view of the analysis chamber.

---

[1] *Low Energy Electron Diffraction*

[2] *Angle Resolved Photo Electron Spectroscopy*

5.1.4. Sample storage and handling

Between the MBE chamber and the STM/preparation chambers there is a sealable section providing room for the storage of up to four sample carriers placed on a vertical lift. This section can be vented separately for sample replacements, and due to its small volume the UHV conditions can be quickly restored by an overnight bake-out.

The specific sample carriers are fully made of tantalum weighing exactly 55 g in order to balance the STM spring system for vibrational decoupling. Their outer shape has been especially designed for convenient transport and mounting within the different sections of the UHV system. In Fig. 5.5 b such a sample carrier is shown in detail.

To safely hold and handle the sample carriers, the different linear and vertical positioners are equipped with special clamps, brackets, or spreaders. The specific mechanisms are sketched in Fig. 5.5 a.

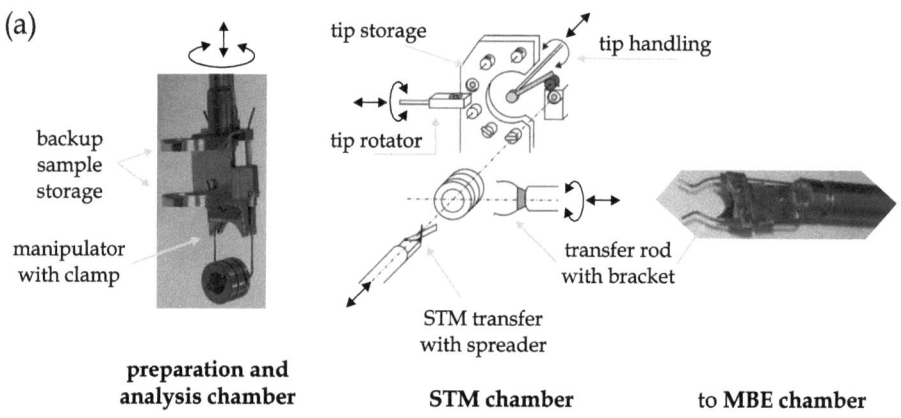

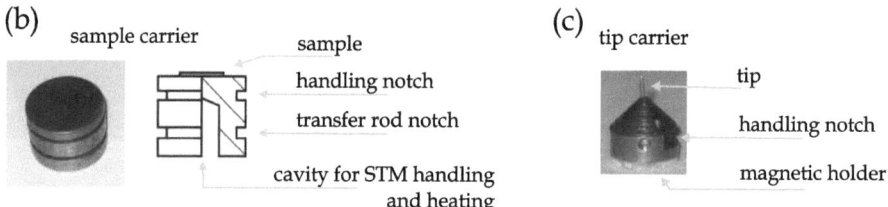

Figure 5.5: (a) Schematic details of the tip handling in the STM setup and the sample handling between the different UHV modules. (b) Photograph and schematic side view of the sample carrier. (c) Side view photograph of the tip carrier.

5.2. Preparation of the experimental setup

There are different aspects concerning the maintenance and preparation of the complete experimental setup.

In case the MBE chamber was vented, the constituent K-cells may have been affected by contaminants, which affect the purity of the molecular beam. To remove such contaminants, the K-cells are outgassed by repeatedly rising the cell temperature slightly above the designated operation temperature, until the observed outgassing pressure does not significantly exceed the residual pressure any more. After that, the MBE setup remains in standby mode, where the effusion cells are kept at an idle cell temperature of $T_{idle} = 80\,°C$ for the As K-cell and cracker and at $T_{idle} = 200\,°C$ for the Ga and In K-cells.

Before any MBE growth cycle starts, the MBE chamber is cooled by filling the cryopanel with ℓ-N_2, in order to reduce the number of residual gas particles. Simultaneously, the K-cells are heated to the designated operational temperature to reduce any effects by leftover contaminants and then put to an idle temperature. This allows fast adjustments of the required partial beam pressures, as needed for specific stages in a growth cycle.

On a regular basis, the material flux from the effusion sources is also controlled by analyzing the BEP. For this purpose, the partial pressure of the molecular beams from the constituent K-cells is observed as a function of the effusion cell temperature. The average BEP data of the constituent effusion cells used in this work are given in Appendix A.

5.3. Preparation of the STM tips

For the STM investigations in this setup platin/iridium (*Pt/Ir*) based STM tips were used, as tungsten based STM tips did not prove sufficient tunneling stability on the GaAs(0 0 1) surface. The first investigations were conducted using self-cut tips made from a 0.5 mm thick $Pt_{0.8}/Ir_{0.2}$ wire, following the techniques of *K. Jacobi* and his co-workers. The wire was simply cut under a steep angle by a side cutter, which was pulled away in the moment of the cut in order to form a tip of sufficient sharpness. However, the reproducibility of such tips highly depends on the individual technique. Moreover, the self-cut tips often did not demonstrate a reliable tunneling stability in STM measurements.

Therefore, the STM results presented in this work were conducted by using electrochemically etched $Pt_{0.8}Ir_{0.2}$ tips, commercially manufactured by *Agilent Technologies*. Such tips exhibit a rather roundly shaped cone with a sharp tip apex in front. Cut to a length of about 10 mm, the tips are then mounted into a tip carrier (Fig. 5.5 c) and stored in the tip storage in the STM chamber (Fig. 5.5 a). Prior to the first use, the tips are cleaned from oxides by electron beam heating at typical values of $V = 250\,V$ and $I = 2.5\,mA$ for a time $t = 30\,s$. This procedure can also be repeated to refresh the tips after intense use.

5.4. Preparation of the sample substrate

The basic substrate material for the experiments in this work was a typical Si-doped n-type GaAs(0 0 1) wafer with a dopant concentration of about $5 \cdot 10^{17}\,\text{cm}^{-3}$, oriented by $\pm 0.1°$, with a thickness of 500 μm and a diameter of 2 inch. From this wafer sections with a size of about $10\times 7\,\text{mm}^2$ were cut to serve as the growth samples. These sections were then adhered to the sample carriers using liquid indium as the adhesive. At a temperature of about 200 °C, a small droplet of indium on top of the sample carrier melts into a liquid film so that the backside of the sample can be adhered onto that film, and is carefully adjusted until it is firmly held by molecular adhesion. Redundant In material is removed in order to avoid the contamination of the sample surface. Before placement into the UHV sample storage, sample carriers are rinsed with pure ethanol to remove coarse impurities and dried by nitrogen gas afterwards.

Shortly before growth samples are treated by argon ion sputter etching (cf. also Sect. 5.1.3). An Ar gas partial pressure of typically $1 \cdot 10^{-4}$ mbar was used. Accelerated by an electric field to 1.0 keV, the Ar^+ ions traverse to the sample surface creating a typical ion current $I_{Ar^+} \approx 10\,\mu\text{A}$. The impinging Ar^+ ions then remove surface material, mainly surface adsorbates and oxides. The typical duration times are 60–80 min.

Following that, the sample is transferred onto the heating position in the MBE chamber for annealing. Typical annealing temperatures for GaAs are $\geq 580°\text{C}$, to desorb remaining surface oxides [174]. The thermal activation also supports the stabilization and re-crystallization of sputtering-induced damages of the yet highly rough crystal surface. As arsenic has a very high vapor pressure, it is essential to provide an adequate background pressure of gaseous arsenic when heating As containing materials like GaAs. At sample temperatures above 327°C, As_2 is significantly desorbed from the sample due to the dissociation of GaAs at the surface [175]. Consequently, the As BEP must compensate the surface losses of arsenic, otherwise the remaining gallium would start to assemble out of the crystal order until it forms little droplets, leaving the sample surface unusable for further crystal growth (Fig. 5.6).

Figure 5.6: GaAs(0 0 1) sample with Ga droplets, which evolved during annealing under insufficient As ambient pressure.

After 60–80 min of annealing at an As BEP of $BEP_{As} = 7.5 \cdot 10^{-6}$ mbar the surface structure stabilizes, which can be observed by a sharpening of the reflexes in the RHEED patterns. This cycle of *ion bombardment and annealing* (IBA) is repeated regularly to refresh samples before growth.

5.5. Calibration of the sample temperature

The setup of an external optical pyrometer facing the sample through a chamber window unfortunately has a disadvantage. With the increasing number and duration of growth cycles, excess vapor material — especially the highly reactive As — condensates at that chamber window, reducing its optical transmissibility. This leads to an increasing offset between the measured temperature at the optical pyrometer and the actual substrate temperature T_S. To account for this, the relation between filament current I_F and the substrate temperature T_S has been determined, which can be found in Appendix C (Fig. C.1). The pyrometer window is cleaned regularly during chamber venting procedures allowing to update this calibration data. Additionally, at the beginning of each growth process, the transition between the $\beta 2(2 \times 4)$ and the $c(4 \times 4)$ reconstruction on the GaAs(0 0 1) surface between 490–510 °C [66] was monitored by RHEED and used for a further temperature control.

In this work, all denoted substrate temperatures T_S have been corrected using these calibration data. The accuracy of the denoted values for T_S is subject to an estimated error of about ± 15 °C.

Part II.

Results and discussion

6. Homoepitaxial growth on GaAs(0 0 1)

6.1. Introduction

Uniformity and reproducibility of results is the greatest demand for the epitaxial growth of thin films and nanostructures in respect of application. Roughness of the substrate surface can easily cause growth defects, that deteriorate the electronic properties of such nanostructures, as they are of very small size. In the InAs/GaAs QD system, the WL is only about 2–3 atomic layers thick, yet it is essential for the transport of charge carriers. A rough growth surface would easily disrupt the continuity of this WL and thereby hinder the mobility of charge carriers, which in worst case could lead to the optoelectronic inactivity of this device. Thus, a highly flat substrate surface with broad terraces of ideally only monoatomic height is demanded.

The standard procedure for sample preparation by IBA treatment (cf. Sect. 5.4) does not suffice to provide such a surface, since the ion bombardment more likely even intensifies the surface roughness. Figure 6.1 shows an STM image of the GaAs(0 0 1) surface after IBA treatment. The surface locally exhibits a stable c(4×4) reconstruction, but yet a very high roughness of several atomic layers. There are many clusters of adsorbed excess material, mostly As (marked exemplarily by blue ovals), that was not incorporated into the surface during the annealing process under As_2 pressure or the quenching afterwards. During the annealing, the surface atoms probably suffer from adequate mobility to sufficiently flatten the surface. The highly damaged surface thus provides many attractive sites for material adsorption, especially lattice defects and step edges, where the strain of mismatched clusters can be efficiently relieved. This effect may lead to a structural disorder especially at step edges (exemplarily marked by magenta ovals).

The incorporation of As into the GaAs crystal structure (i.e. chemisorption) requires some activation energy and moreover an adequate partner site at the crystal surface. Yet, during the annealing process only As is offered to compensate for the As vapor losses from the surface, but Ga is not offered. Under these conditions the chemisorption of excess As into the GaAs surface is unlikely as the As_2 sticking coefficient α_{As_2} becomes zero when no excess Ga is available [176]. The crystallization of further GaAs can be induced by offering both species Ga and As at an adequate substrate temperature T_S (*homoepitaxy*). In order to compensate most surface damages and obtain a highly flat and clean GaAs surface a

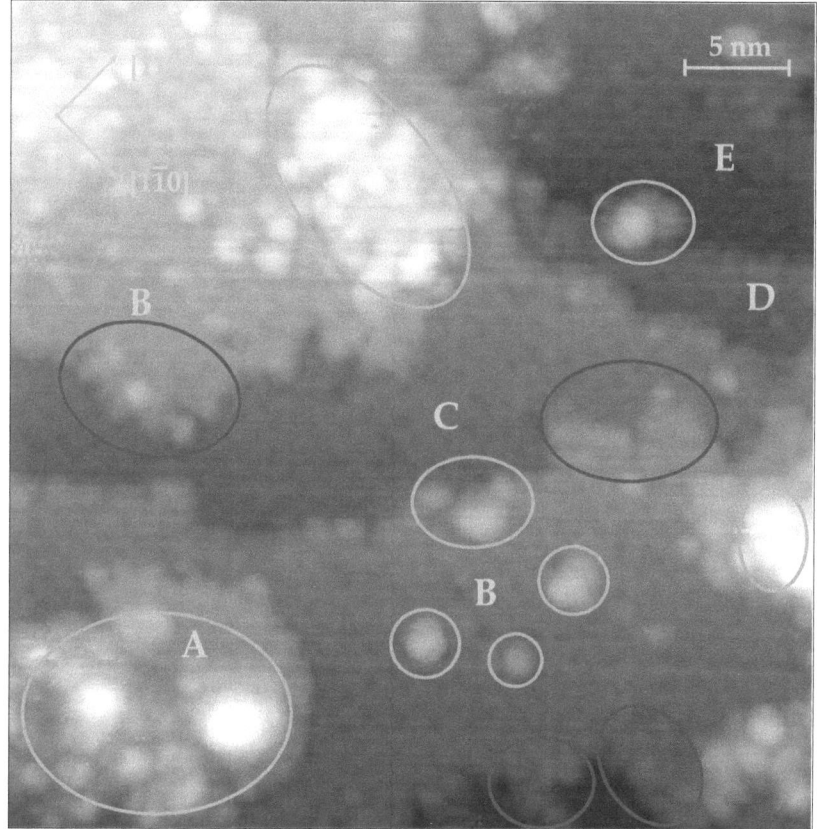

Figure 6.1: Filled state STM image ($V_T = -2.5\,V$, $I_T = 0.15\,nA$) of the GaAs(0 0 1)-c(4×4) reconstructed surface after IBA treatment. The apparent atomic terraces are marked by letters A to E. Clusters of excess material are marked exemplarily by blue ovals. Magenta ovals mark disordering effects at step edges.

homoepitaxial layer of sufficient thickness (*buffer layer*) must be grown on the IBA treated substrate.

6.2. Experimental details

6.2.1. Sample preparation

After IBA treatment the sample surface was observed by RHEED and T_S was slowly reduced, while the As$_2$ partial pressure was maintained at $BEP_{As} = 3 \cdot 10^{-6}$ mbar. Initially the RHEED patterns showed (2×4) periodicity, which corresponds to the $\beta 2(2\times 4)$ reconstructed

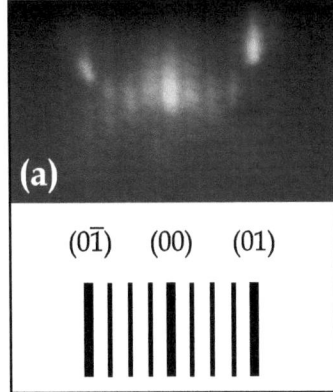 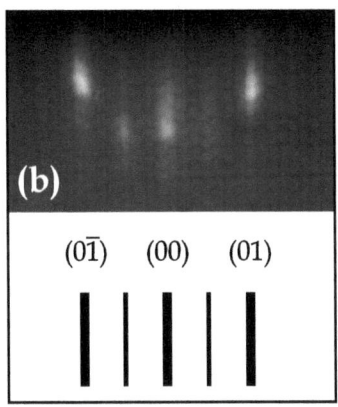

Figure 6.2: *RHEED patterns of the GaAs(0 0 1) surface taken before and after the periodicity transition from (2×4) to c(4×4) symmetry with incident beam along the [$\bar{1}$ 1 0] direction; (a) shows RHEED patterns of the β2(2×4) reconstruction at $T_S \geq 515\,°C$ and (b) shows RHEED patterns of the c(4×4) reconstruction at $T_S \leq 485\,°C$.*

surface. At $T_S \approx 500\,°C$ the transition of the surface reconstruction from the β2(2×4) reconstruction to the c(4×4) reconstruction was observed, used to confirm the calibration of T_S. The corresponding RHEED patterns are shown in Fig. 6.2.

For buffer layer growth a substrate temperature $T_S = 515\,°C$ was chosen, typical for GaAs homoepitaxy on the β2(2×4) surface, as As induced structural growth defects can be efficiently reduced in this temperature range [177]. An appropriate As/Ga flux ratio of about 25 was chosen to ensure As rich growth conditions. The temperature of the As K-cell was set to $T_{As} = 475\,°C$, corresponding to a partial pressure $BEP_{As} = 3.5 \cdot 10^{-6}\,\text{mbar}$, and the temperature of the Ga K-cell was set to $T_{Ga} = 1010\,°C$, corresponding to a partial pressure $BEP_{Ga} = 1.4 \cdot 10^{-7}\,\text{mbar}$ (see Appendix A, Figs. A.1, A.2). Then the resulting V/III flux ratio can be derived from

$$\frac{BEP_{As}}{BEP_{Ga}} = \frac{3.5 \cdot 10^{-6}\,\text{mbar}}{1.4 \cdot 10^{-7}\,\text{mbar}} = 25. \tag{6.1}$$

The temperature of the As cracker was maintained at $T_{Cr} = 985\,°C$ whenever the As K-cell was in operation.

All grown samples presented in this work were grown with a buffer layer for about 60 min under these growth conditions.

6.2.2. Determination of the growth rate

During growth, the intensity of the RHEED specular beam was monitored to verify the growth rate. For this purpose a photodiode was fixed at the specular spot position on the RHEED screen and the photocurrent was measured over time. Yet, the focusing of this setup was not optimal and scattering light and small shifts of the spot position roughened

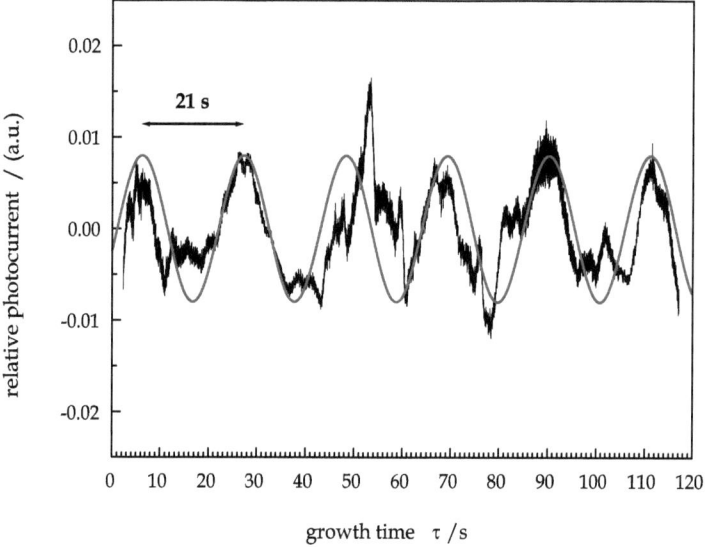

Figure 6.3: Normalized photocurrent data profile of a photodiode observing the intensity fluctuations of the RHEED specular spot during GaAs homoepitaxy. To adapt to the high noise level, the periodic fluctuations were fit a by sine function (red). The oscillation period of this sine function then is 21 s.

the measured signal. A typical set of the photocurrent data for GaAs homoepitaxy is shown in Fig. 6.3. The average duration between two intensity peaks is 21 ± 1 s, corresponding to the time t_m during which one ML of evaporant material is deposited. The typical growth rate for GaAs then is $1/t_m = 0.05$ ML/s. Hence, a growth time of 60 min for the buffer layer corresponds to 180 ML or a thickness of about 50 nm, considering that on the GaAs(0 0 1) surface 1 ML = $a/2$ = 0.283 nm.

The typical growth rate derived from monitoring the RHEED specular spot may be compared to the theoretical growth rate basically derived from the rate of particle impingement j (Eq. 3.1) and the BEP measured at the sample growth position. During highly As-rich growth conditions, the GaAs growth rate is determined solely by the number of available Ga atoms, as the sticking coefficient of Ga atoms α_{Ga} can be assumed as unity [176]. Thus it is adequate to consider only BEP_{Ga} to calculate the BEP-derived growth rate, which yields about 0.02 ML/s then. This value deviates by a factor of about 2.5 from the measured growth rate, which must be ascribed mainly to the inaccuracy of BEP measurements using an ion gauge that is actually standardized for detecting hydrogen and nitrogen gas molecules. However, the BEP-derived growth rate still is in the expected magnitude, supporting the experimental findings.

6.3. The GaAs(0 0 1)-c(4×4) reconstructed surface

6.3.1. Sample growth parameters

The growth scheme of the following sample is illustrated in Fig. 6.4. During the complete procedure the sample was exposed to an ambient As_2 pressure of $BEP_{As} = 3.5 \cdot 10^{-6}$ mbar. After buffer layer growth the sample was annealed for about 20 min at $T_S = 515\,°C$ to obtain a stable and mostly undisturbed $\beta 2(2 \times 4)$ reconstructed surface, while the sample surface was continuously monitored with RHEED. A sharpening of the RHEED patterns indicates a sufficient surface stabilization. Then the sample temperature was reduced to $T_S = 460\,°C$ and the sample was annealed for a time $t = 10$ min while the formation of the c(4×4) surface reconstruction was observed by RHEED. The transition between both reconstructions occurring at 500 °C [66] was used to calibrate T_S.

To preserve the surface for STM studies, the sample is rapidly quenched in less than 2 min below a temperature of $T_S < 327\,°C$ [175] at which the As K-cell needle valve is closed to reduce the ambient As_2 pressure. The sample is then transferred into the STM for analysis.

6.3.2. STM results

Overview

The STM image in Fig. 6.5a provides an overview over the bare GaAs(0 0 1)-c(4×4) surface with atomic resolution. The typical brick-like alignment of the bright appearing As triple dimers can be clearly observed. In filled-state STM images the three As dimers are usually appear as one bright brick. The corresponding c(4×4) surface unit cell is marked at Pos. ①. Spots of bright contrast (marked exemplarily by blue ovals) result from excess atoms, mostly As, that were merely adsorbed at the surface during quenching.

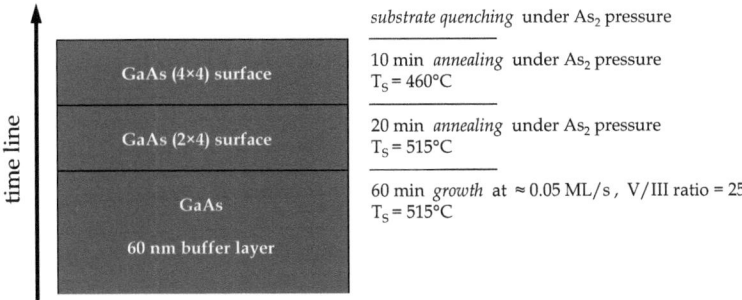

Figure 6.4: Sample growth scheme for the preparation of the GaAs(0 0 1)-c(4×4) surface. The time line in z-direction indicates the chronological progression as well as the resulting sample structure.

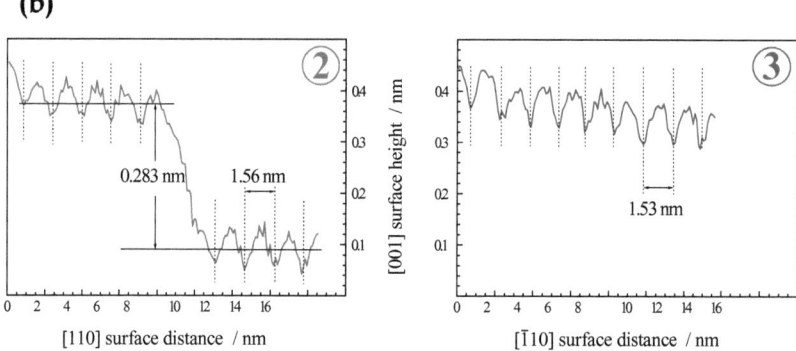

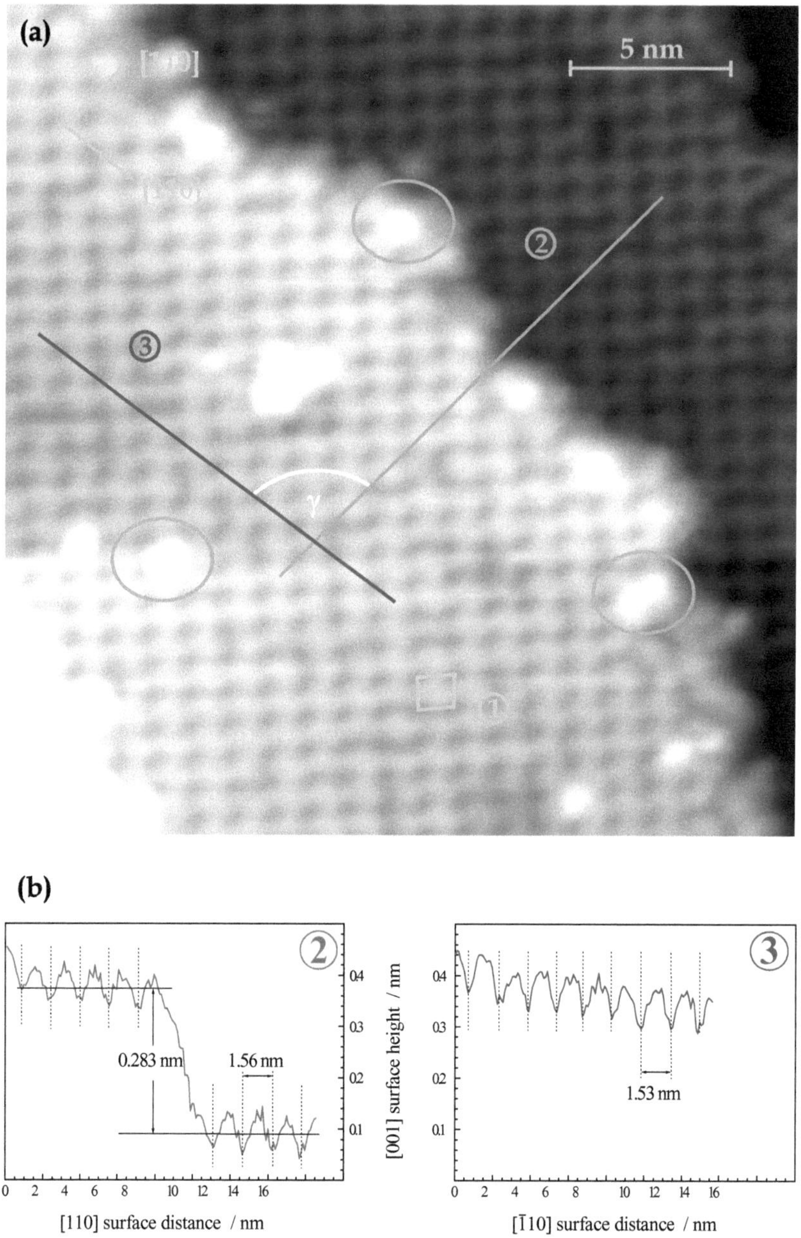

Figure 6.5: (a) Filled state STM image ($V_T = -2.7\,V$, $I_T = 0.2\,nA$) of the GaAs(0 0 1)-c(4×4) reconstructed surface after buffer layer growth and annealing. Blue ovals exemplarily mark the spots of bright contrast caused by excess As adsorbed during quenching. The c(4×4) surface unit cell ① is marked by a yellow box. The location of the height profile lines is marked by ② and ③, the corresponding data is presented in (b).

6.3. THE GAAS(0 0 1)-C(4×4) RECONSTRUCTED SURFACE

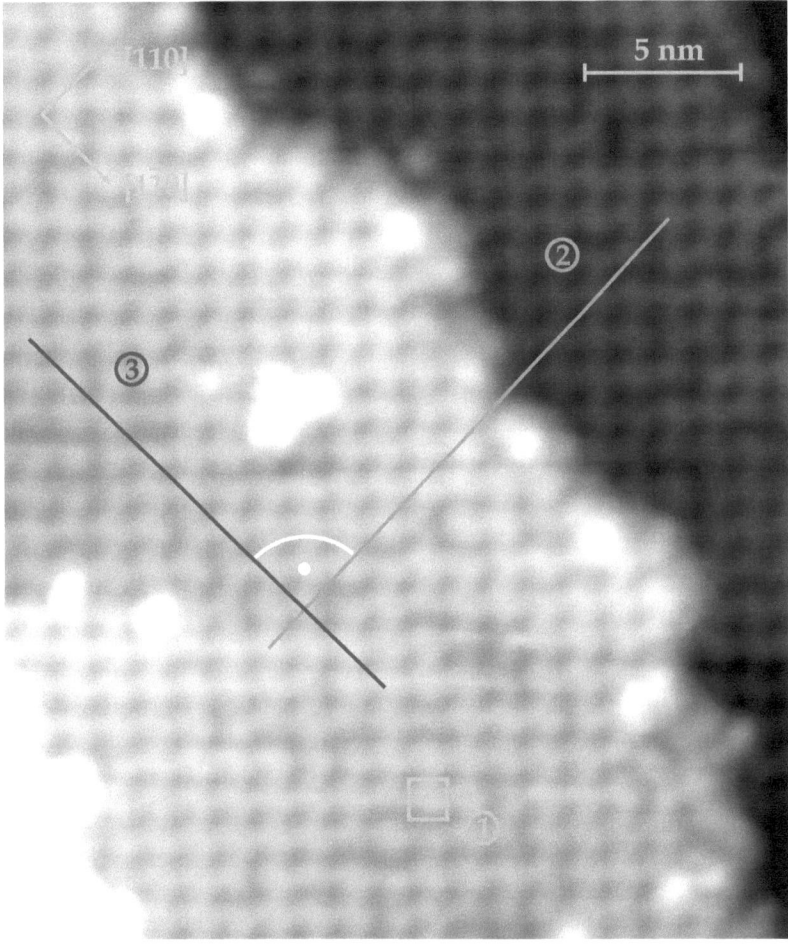

Figure 6.6: Filled state STM image as in Fig. 6.5a ($V_T = -2.7$ V, $I_T = 0.2$ nA) after correcting the lateral deformation caused by drift effects.

The buffer layer growth has smoothened the surface, which exhibits large flat terraces, separated by steps of monolayer height. From a height profile at Pos. ② in Fig. 6.5a the step height can be derived. The corresponding data is shown in Fig. 6.5b. The measured step height between terraces is $h = 0.283 \pm 0.008$ nm, which clearly matches the value of a monolayer height on the GaAs(0 0 1) surface.

Furthermore, from the periodic oscillations in the height profile the distance between two As vacant positions (*hollow sites*) can be measured. In [1 1 0] direction (height profile ②) this measured distance is $d_{[110]} = 1.56\pm0.01$ nm. In [$\bar{1}$ 1 0] direction (height profile ③) the measured distance between hollow sites is $d_{[\bar{1}10]} = 1.53\pm0.03$ nm. The expected distance derived from the geometry of the surface reconstruction is 1.60 nm corresponding to the length of four aligned (1×1) crystal surface unit cells.

This lateral offset in the STM data is most probably due to drift effects during scanning. As the tube piezo scans the surface line by line, the scanning speed in line is fast and continuous compared to the perpendicular direction, where the surface is scanned more slowly. In STM operated at room temperature thermal stability is difficult to maintain and thus temperature fluctuations in the order of 1 K/h within the setup are often encountered. These fluctuations cause shifts in the lateral alignment of tip and sample, that can be significant on the atomic scale. Yet, the influence of that shift is much higher in the slow scan direction perpendicular to the scan lines than it is along the scan line and can result in a deformation of the STM image.

This deformation is evident in Fig. 6.5 a considering the nominally right angle γ enclosed between both profile lines ② and ③, which appears clearly larger in this STM image ($\gamma = 100°$). However, if there is sufficient structural data that can be retrieved from the STM image, e.g. atomic resolution, and the geometric properties of the expected structure is known, the deformation of the STM image data can be corrected. The resulting image after deformation correction is shown in Fig. 6.6. In the following, all STM images providing sufficient data are presented after using this correction procedure.

Structural details

The STM image in Fig. 6.7 provides a detailed look into the atomically resolved surface structure of the GaAs-c(4×4) surface. The overlaid structure model nicely confirms the observed filled state contrast of the As triple dimers of the surface unit cells well. In the structural model these top-most As dimers are illustrated by pairs of large white circles, gray filled circles correspond to bulk As atoms and blue filled circles correspond to bulk Ga atoms. The size and location of a surface unit cell is marked in orange by a dashed square.

Despite the well ordered appearance of the c(4×4) surface observed on a large scale, a detailed look into the local surface structure at the scale of some surface unit cells reveals that specific local structural defects are quite likely to occur. Such defects are exemplarily shown in Fig. 6.8.

Most common is the formation of local Ga–As heterodimers rather than As–As dimers (Fig. 6.8 a), a competing process during growth or quenching mainly influenced by the As partial pressure [56]. The replacement of further As atoms to form Ga–Ga dimers is also possible, yet an energy barrier is delaying such a process [81]. From only filled state images the conclusion cannot be safely drawn, that the missing contrast at a dimer site is not due

6.3. THE GAAS(0 0 1)-C(4×4) RECONSTRUCTED SURFACE

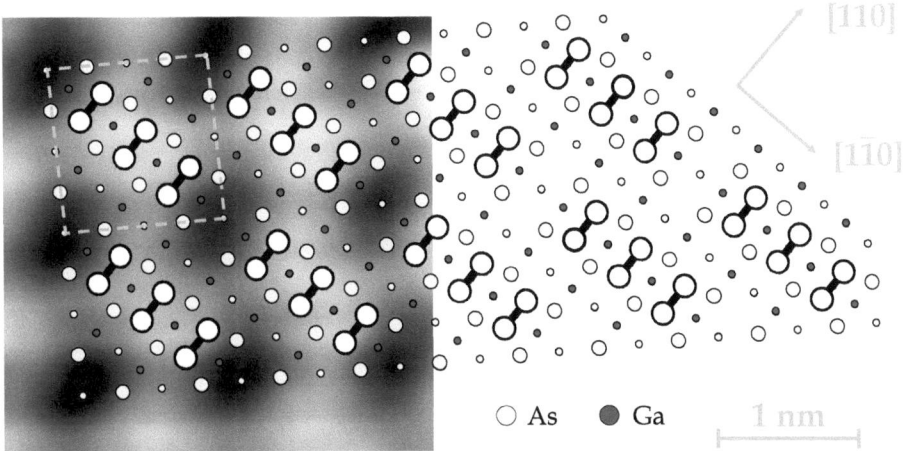

Figure 6.7: *Filled state STM image ($V_T = -2.7$ V, $I_T = 0.2$ nA) of the GaAs(0 0 1)-c(4×4) surface structure in detail. The color scheme of the overlaid structural model is blue for Ga atoms, gray for As atoms, and white for the top-most As dimer atoms.*

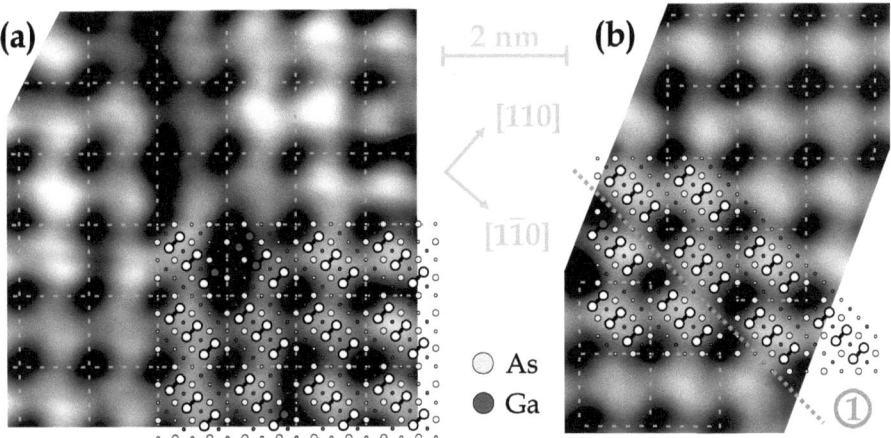

Figure 6.8: *Filled state STM images ($V_T = -2.7$ V, $I_T = 0.2$ nA) of the GaAs(0 0 1)-c(4×4) surface structure exhibiting typical structural defects that were observed in detail; (a) the formation of Ga–As heterodimers and Ga–Ga dimers, and (b) the formation of dislocations of the surface unit cells along a dislocation line (marked by a dotted blue line ①). In both images the location of the surface unit cells is indicated by a mesh of dashed orange lines. The color scheme of the overlaid structural model is blue for Ga atoms, gray for As atoms, and white for the top-most As dimer atoms.*

to completely missing atoms/dimers. However this is presumed to be energetically less favorable compared to a Ga atom/dimer occupation.

The formation of lateral dislocations of the surface reconstruction is shown in Fig. 6.8 b. In this case the brick-like structure is shifted along a dislocation line in [$\bar{1}$ 1 0] direction ① by ¼ of a surface unit cell, as the overlaid structural model reveals.

6.3.3. Summary

The preparation of the GaAs-c(4×4) surface by MBE results in a flat, well-ordered, and stable surface with large terraces separated by steps of monolayer height. Some contamination by adsorbed excess material during quenching is apparently inevitable. Locally, the surface exhibits some structural defects in the atomic configuration, mostly replacements of atomic species and lateral dislocations. The high reproducibility of structural results allows the calibration of the STM instrument, particularly with regard to further investigations of possibly unknown structures.

6.4. The GaAs(0 0 1)-$\beta 2(2 \times 4)$ reconstructed surface

6.4.1. Sample growth parameters

The growth scheme of the following sample is illustrated in Fig. 6.9. During the complete procedure the sample was exposed to an ambient As_2 pressure of $BEP_{As} = 3.5 \cdot 10^{-6}$ mbar. After buffer layer growth the sample was kept at $T_S = 515\,^\circ$C and the sample was annealed for $t = 10$ min. During annealing the stabilization of the $\beta 2(2 \times 4)$ surface reconstruction was observed by RHEED, indicated by a sharpening of the reflex patterns.

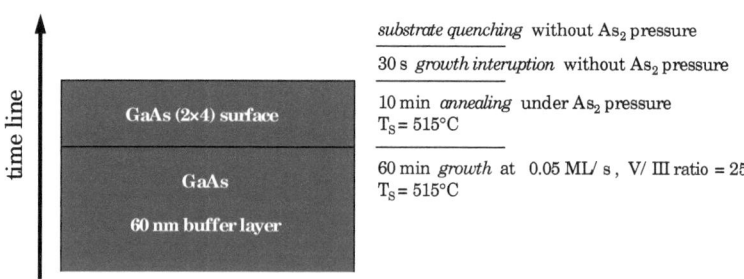

Figure 6.9: *Sample growth scheme for the preparation of the GaAs(0 0 1)-$\beta 2(2 \times 4)$ surface. The time line in z-direction indicates the chronological progression as well as the resulting sample structure.*

To preserve this surface for STM studies, first the As atmosphere was reduced by closing the valve from the As K-cell, while the sample was kept at $T_S = 515\,°C$ for a growth interruption of about 30 s. Then the sample was rapidly quenched under RHEED observation to ensure that the $\beta2(2\times4)$ does not reconstruct to the more As-rich (4×4) configuration during this process. Finally the sample was transferred into the STM for analysis.

6.4.2. STM results

Overview

The STM image in Fig. 6.10 a provides an overview over the bare GaAs(0 0 1)-$\beta2(2\times4)$ surface. The typical structure of parallel As dimer rows along the $[\bar{1}\,1\,0]$ direction is clearly observed. Local spots of bright contrast are most likely caused by excess As material that was adsorbed at the surface during quenching. Exemplary spots are marked by blue ovals.

Similar to the c(4×4) surface preparation, the buffer layer growth has smoothened the surface, which exhibits flat terraces with steps of monolayer height. The data from the height profile at Pos. ① is presented in Fig. 6.10 b. The measured step height here is $h = 0.265\pm0.012$ nm, which even considering the error margins is slightly less than the theoretical value of 0.283 nm. The difference might be due to different tip tunneling conditions caused by step charging as well as surface defects.

Structural details

A more detailed view of the surface structure is given in Fig. 6.11 a. The rows of double As top dimers and the trenches, both characterizing this surface reconstruction, are evident. The position of a surface unit cell is marked by a yellow rectangle ①. From a height profile ② in [1 1 0] direction the distance between the surface trenches can be derived, which corresponds to the distance between two surface unit cells. The corresponding height profile data is presented in Fig. 6.11 b. The average measured distance between trenches is $d_{[110]} = 1.81\pm0.02$ nm, which is slightly more than the nominal value of 1.60 nm corresponding to the length of four aligned (1×1) crystal surface cells. This difference might be caused by tip-length effects or resulting from fast scanning.

In the STM studies of this work tunneling on the $\beta2(2\times4)$ reconstructed surface with sufficient atomic resolution was generally difficult. However, as the image in Fig. 6.11 c reveals, local atomic resolution even of the trench dimer was evidently observed. The corresponding surface unit cell is marked by a dotted yellow rectangle. The overlaid structural model confirms the location of the top-most pair of As dimers, followed by a trench, where also the small contrast of the lower trench dimers is observed.

However, specific details on surface defects could not be revealed, as the atomic resolution in the STM images did not suffice to provide reliable data.

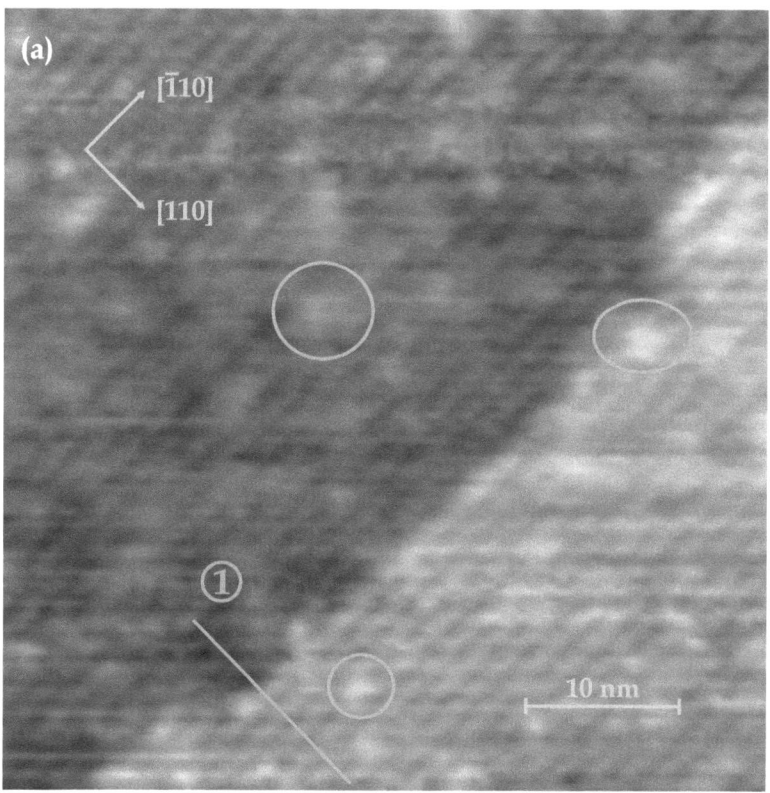

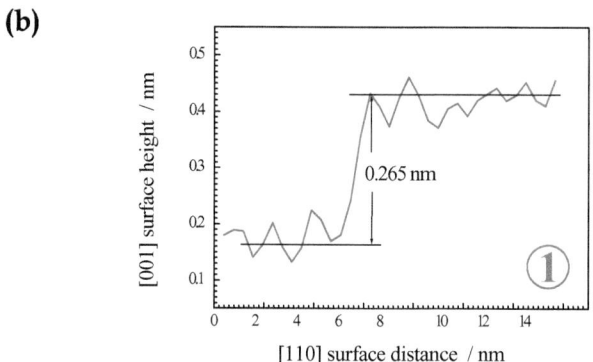

Figure 6.10: (a) Filled state STM image ($V_T = -3.0\,V$, $I_T = 0.5\,nA$) of the GaAs(001)-$\beta 2(2\times 4)$ reconstructed surface after buffer layer growth and annealing. Blue ovals exemplarily mark the spots of bright contrast caused by excess As material adsorbed during quenching. From a height profile ① the step height between terraces is derived. The corresponding profile data is presented in (b).

6.4. THE GaAs(001)-β2(2×4) RECONSTRUCTED SURFACE

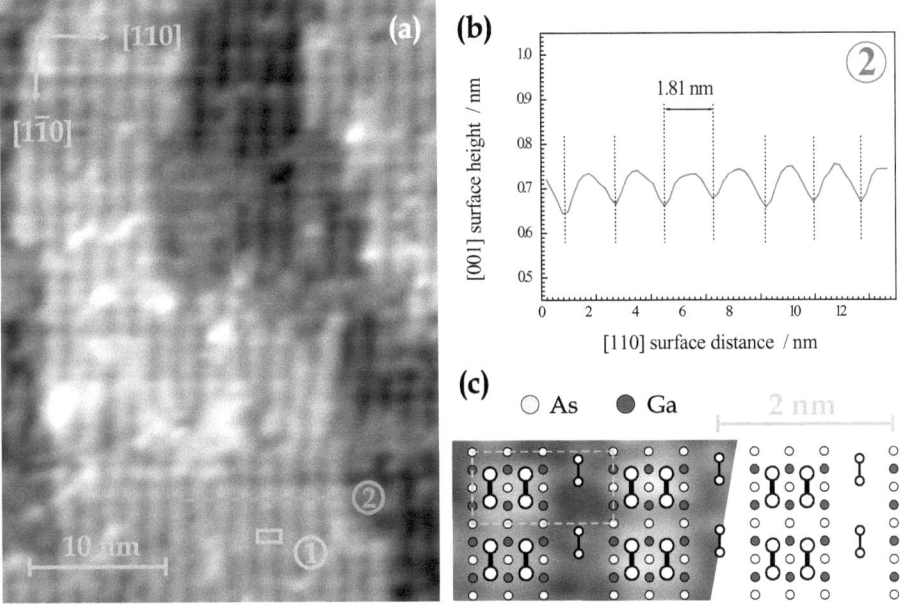

Figure 6.11: ($V_T = -3.0\,V$, $I_T = 0.6\,nA$) (a) Filled state STM image of the GaAs(001)-β2(2×4) reconstructed surface. The corresponding surface unit cell is marked by a yellow box ①. The data from the height profile ② illustrated in (b) shows the corrugation of the typical As dimer chains. In (c) the structural model is laid over an atomically resolved image detail.

6.4.3. Summary

The preparation of the GaAs(001)-β2(2×4) surface by MBE results in a smooth and stable surface, with large terraces separated by steps of monolayer height. The surface appears largely well ordered, as far as the smeared atomic resolution allowed to observe. The smeared atomic resolution also hindered from verifying structural defects in the surface reconstruction. STM studies on the β2(2×4) reconstructed surface unfortunately suffered from insufficient atomic resolution at most preparations. Nevertheless, the detailed structure of the As dimer alignment and the trench dimer could be verified.

7. InAs thin film growth on GaAs(0 0 1)-c(4×4)

7.1. Introduction

Though the growth of InAs/GaAs QDs has become a common technique for applications, parts of this growth process are yet not understood fully in detail. While there is a good agreement among reports on the formation of the QDs themselves, i.e. the SK-growth mode and conditions for the 2D→3D transition, details on the complex evolution of the first atomic layers of the deposited InAs on the GaAs(0 0 1) substrate, the WL, and its detailed structure before and after the 3D transition still remain unclear.

The reported surface reconstructions for the InAs WL vary from (1×3) [71, 103] and (2×3) [104, 105] derived from diffraction methods to (4×3) [106, 107] and (4×6) [178] derived from STM studies. The diffraction methods may not be able to resolve the surface periodicity appropriate, if there is a significant occurrence of surface domains or surface defects and in result missing long-range order. The STM reports unfortunately suffer from inadequate atomic resolution to verify a structural model. Differing growth conditions may also contribute to these seemingly inconsistent findings.

7.2. Experimental details

7.2.1. Sample preparation

In order to systematically investigate the evolution of InAs thin films on GaAs(0 0 1), the sample growth principally followed the same growth scheme as for the homoepitaxy on GaAs(0 0 1), but with increasing amounts of additionally deposited InAs. This sample growth scheme is illustrated in Fig. 7.1. During the complete procedure the sample was exposed to an ambient As$_2$ pressure of $BEP_{As} = 3.5 \cdot 10^{-6}$ mbar.

After buffer layer growth and annealing, the sample temperature was reduced to $T_S = 460\,°C$ to obtain the c(4×4) reconstructed surface. For stabilization the sample was annealed for another 10 min while monitored by RHEED. In the next growth stage InAs was deposited in increasing amounts at an As/In flux ratio of about 200. After the InAs deposition, the sample surface was rapidly quenched in less than 1 min below 327 °C, where the ambient As_2 pressure was reduced. During this procedure, changes in the RHEED patterns of the grown surface structure were never observed, which indicates that no structural changes of the surface took place during quenching and that the surface is in the same structural stage as immediately after growth.

After quenching the sample was directly transferred into the STM chamber for structural analysis.

7.2.2. Determination of the growth rate

The InAs was deposited at a very low In flux ($BEP_{In} \approx 1.9 \cdot 10^{-8}$ mbar) corresponding to a growth rate of about 0.007±0.002 ML/s. Such a slow growth rate ensures the controlled deposition of very little InAs amounts in the submonolayer range [179].

However, due to the resulting small film thickness, the growth rate cannot be sufficiently determined using the oscillation of the RHEED specular spot, as it was used to determine the GaAs growth rate. Furthermore, technical reasons required the shutter of the In K-cell to remain open, and the material deposition was merely controlled by the heating procedure of the In crucible, the slightly varying K-cell temperature T_{In} during growth and the deposition time τ. Thus, the InAs growth rate had to be determined using recursive methods and was calibrated afterwards using the resulting RHEED and STM data. A more detailed discussion on this calibration is given in Appendix B.

7.3. InAs thin films at submonolayer coverages

7.3.1. Sample growth parameters

Three samples with increasing InAs coverages starting from 0.09 ML to 0.56 ML were investigated, grown according to the general sample growth scheme. The InAs deposition time for each of these samples and the resulting amount of deposited material, as derived from the InAs growth function in Appendix B, is given in Tab. 7.1. During growth of each sample, no change of the initial c(4×4) periodicity was observed by RHEED.

7.3.2. STM results

Overview

The STM images in Figs. 7.2, 7.3, and 7.4 show overview images of the GaAs(0 0 1)-c(4×4) reconstructed surface after the deposition of increasing amounts of InAs material. On a

7.3. InAs thin films at submonolayer coverages

large scale the surface of each sample appears very smooth and clean with large flat terraces separated by steps of monoatomic height.

Evaluating the images on a smaller scale, the global distribution of the InAs on the surface can be observed from the corresponding signatures with brighter contrast. For a very low material amount of about 0.09 ML of InAs (Fig. 7.2 b), these signatures are found at the fringes of areas with clean GaAs-c(4×4) reconstruction, aligning in an apparently random mesh-like structure. The number of these InAs signatures then increases proportionally with the deposited amount of InAs. For an InAs coverage of about 0.30 ML, only some small domains of a pure GaAs-c(4×4) reconstructed surface are left (Fig. 7.3 b). The mesh-like structure of the InAs signatures has dissolved, yet it still appears as there is a preferential alignment along the [$\bar{1}$ 1 0] direction. Finally, for an amount of about 0.56 ML of deposited InAs, the signatures disperse evenly over the surface (Fig. 7.4 b). Significant domains of pure GaAs-c(4×4) were not observed any more.

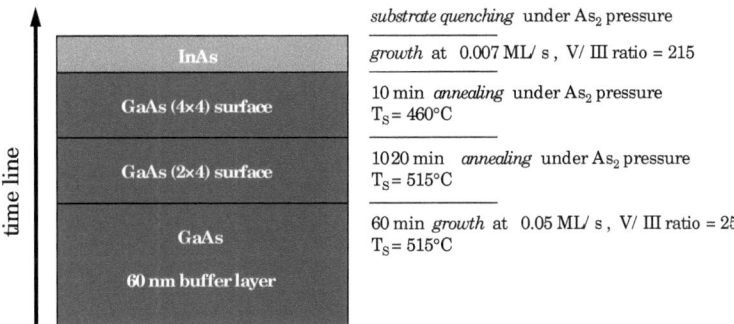

Figure 7.1: *General sample growth scheme for the preparation of InAs thin films on the GaAs(0 0 1)-c(4×4) surface. In the last growth stage, i.e. the InAs deposition, the amount of deposited InAs was systematically increased.*

	sample A	sample B	sample C
deposition time τ	20 s	52 s	80 s
calculated InAs deposition derived from the growth function	0.09±0.03 ML	0.30±0.04 ML	0.56±0.05 ML
1st layer coverage of (4×4) unit cells *	10±3%	30±4%	50±5%
estimated total InAs content derived from the STM Data	0.10±0.03 ML	0.30±0.04 ML	0.50±0.05 ML

Table 7.1: *InAs growth parameters for samples A–C grown with InAs submonolayer coverages investigated in the following. The presented data relies on the STM findings and follows the discussion in Appendix B.*
*The surface area of one c(4×4) surface unit cell corresponds to $A = 1.28$ nm^2.

Figure 7.2: Sample [A]: Filled state STM images ($V_T = -2.7\,V$, $I_T = 0.2\,nA$) of the GaAs(0 0 1)-c(4×4) reconstructed surface covered with 0.09 ML of InAs material. The large scale image in (a) exhibits a smooth surface with flat terraces of monoatomic height, the more detailed view in (b) reveals large areas of the GaAs-c(4×4) reconstruction and a mesh-like structure of signatures with bright contrast caused by the deposited InAs.

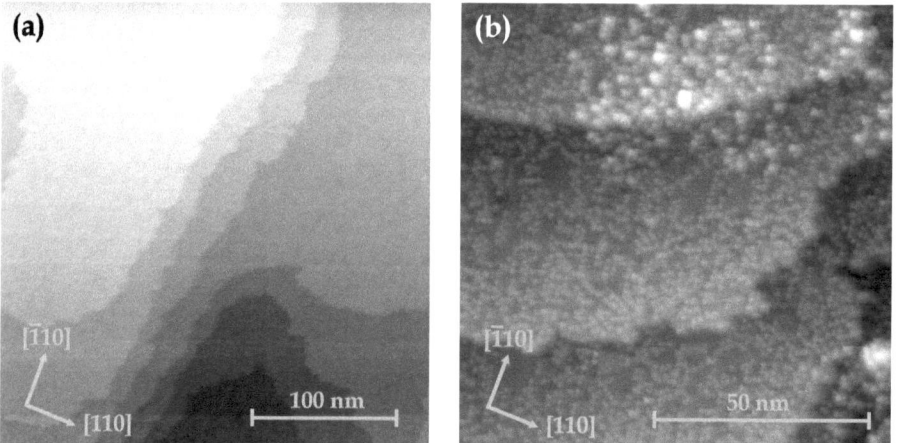

Figure 7.3: Sample [B]: Filled state STM images ($V_T = -2.7\,V$, $I_T = 0.2\,nA$) of the GaAs(0 0 1)-c(4×4) reconstructed surface covered with 0.30 ML of InAs material. In the large scale image (a) the surface appears very smooth with flat terraces of monoatomic height. In a more detailed view (b) the increased coverage with the InAs signatures is clearly observed.

7.3. InAs thin films at submonolayer coverages

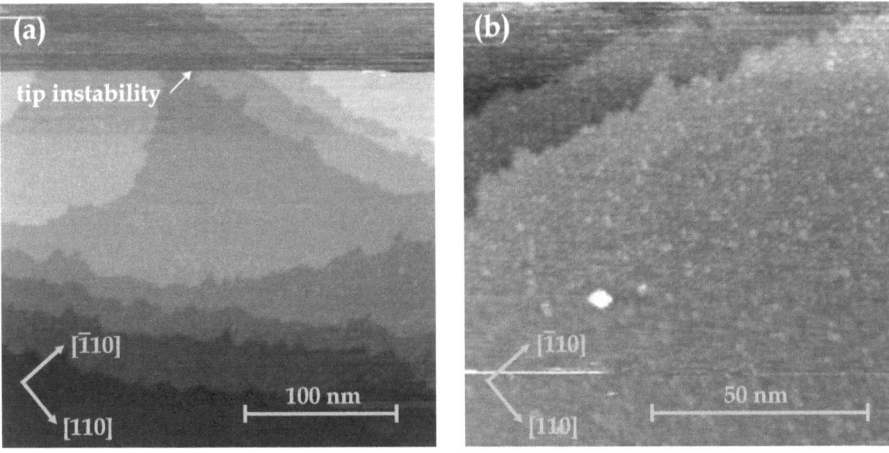

Figure 7.4: Sample [C]: Filled state STM images ($V_T = -2.7\,V$, $I_T = 0.2\,nA$) of the GaAs(0 0 1)-c(4×4) reconstructed surface covered with 0.56 ML of InAs material. Again, the large scale image (a) shows a smooth surface with large terraces. In the more detailed view in (b) the initial c(4×4) is not clearly evident any more, and the observed InAs signatures are now distributed evenly over the surface.

The tunneling stability of the STM investigations clearly worsened with increasing InAs amounts on the surface.

Structural details

Investigating the InAs distribution on the atomic scale, it becomes evident that there are preferential sites for the InAs adsorption (Fig. 7.5 a). The InAs signatures are generally found at the hollow sites of the c(4×4) reconstruction. Yet most of the InAs assembles at hollow sites neighboring to surface dislocations or comparable surface defects of the c(4×4) reconstruction ①. Much less InAs signatures are found at the hollow sites of the undisturbed domains of the bare GaAs-c(4×4) surface ②. The more detailed view in Fig. 7.5 b again shows that the domains of undisturbed GaAs-c(4×4) barely show any InAs signatures, while most InAs is adsorbed near defect sites of the surface reconstruction. Such sites may be missing As top dimers of the c(4×4) reconstruction ③ or periodicity dislocations ④.

The InAs signatures are virtually all of equal size and appear clearly separated from each other. Derived from a typical height profile in Fig. 7.6, the height of the signatures slightly exceeds the height of the topmost As dimers on the surface by about 0.08 nm. The signatures exhibit a typical width of about 0.7 nm in [1 1 0] direction and a typical length of about 1.1 nm in [$\bar{1}$ 1 0] direction.

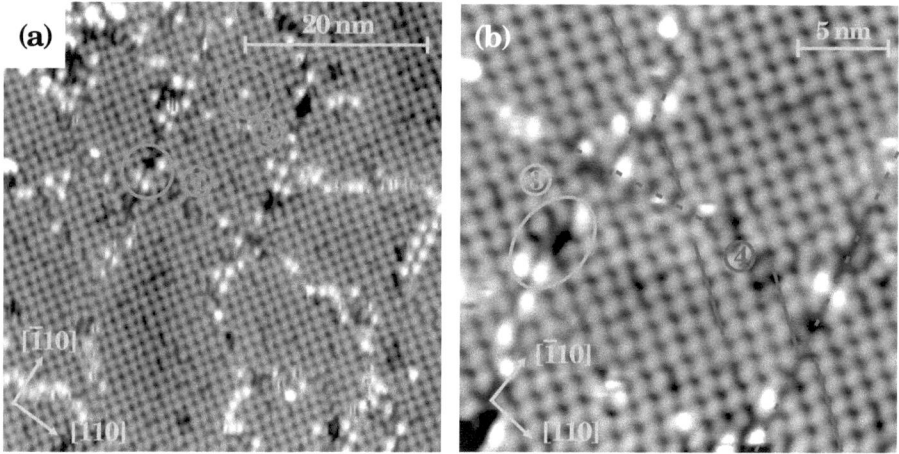

Figure 7.5: Sample [A]: Filled state STM images ($V_T = -2.7\,V$, $I_T = 0.2\,nA$) of the GaAs(0 0 1)-c(4×4) reconstructed surface covered with 0.09 ML of InAs material. The atomically resolved image in (a) illustrates the preferential adsorption sites for the InAs, exemplarily marked by orange circles. Adsorption sites at surface defects are marked by ① and adsorption at an undisturbed c(4×4) hollow site is marked by ②. In a magnified image (b) the preferential adsorption of InAs at surface structure defects ③ and dislocations ④ is evident. The coexisting domains of a nearly undisturbed GaAs-c(4×4) surface mostly remain signature free.

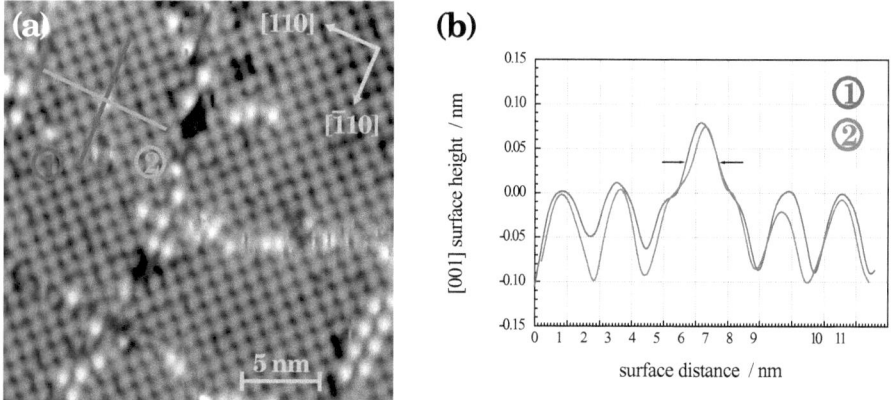

Figure 7.6: Sample [A]:(a) Magnification of a filled state STM image ($V_T = -2.7\,V$, $I_T = 0.2\,nA$) of the sample surface showing the positions of the height profile scans ① and ② across a typical InAs signature. The corresponding data is shown in (b).

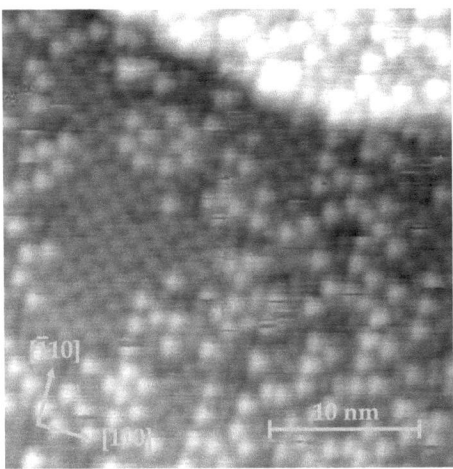

Figure 7.7: Sample B: Filled state STM image ($V_T = -2.7\,V$, $I_T = 0.2\,nA$) of the sample surface covered with 0.30 ML of InAs. As the number of related signatures increases with the amount of deposited InAs material, their size remains unchanged. After populating the hollow sites at surface defects, the signatures increasingly decorate the hollow sites of the undisturbed GaAs(0 0 1)-c(4×4) reconstruction.

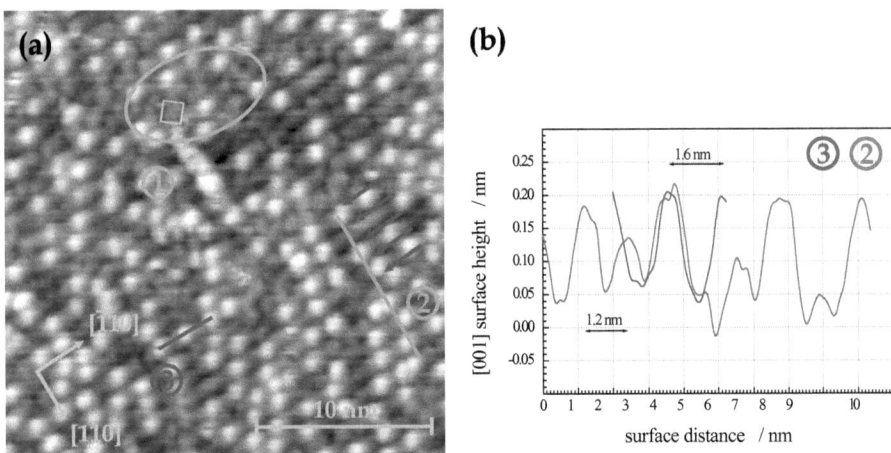

Figure 7.8: Sample C: (a) Filled state STM image ($V_T = -2.7\,V$, $I_T = 0.1\,nA$) of the sample surface covered with 0.56 ML of InAs, showing a nearly uniform distribution of the InAs signatures. Domains of the underlying c(4×4) are only rarely observed (①), the corresponding surface unit cell is marked by a yellow box. The signature alignment is characterized by trenches extending along the [$\bar{1}$ 1 0] direction (marked by magenta arrows). The height profiles ② and ③ indicate the transformation into a more favorable (4×3) surface reconstruction. In (b) the corresponding profile data are given.

On sample Ⓐ they cover about 10% of the surface unit cells, estimated from comparing the possible number of hollow sites to the number of observed signatures, as each c(4×4) surface unit cell is characterized by exactly one hollow site.

As more InAs is deposited, more surface area is covered by the signatures. On sample Ⓑ with 0.30 ML deposited InAs about 30% of surface unit cells are affected (Fig. 7.7). Here, the InAs signatures begin to populate the hollow sites of the undisturbed GaAs-c(4×4) reconstructed domains as well. This effect is intensified if even more InAs is deposited. On sample Ⓒ, where the InAs signatures cover about 50% of the initial c(4×4) surface unit cells, an almost uniform distribution of signatures is observed (Fig. 7.8).

However, the typical size and shape of the InAs signatures remains mostly unchanged on all three investigated samples, seemingly independent from the deposited InAs amount. At surface defects the length of some signatures appears slightly extended up to about 1.6 nm in $[\bar{1}\,1\,0]$ direction.

The STM results show that with increasing InAs coverage, the underlying initial GaAs-c(4×4) reconstruction becomes less observable and surface defect lines of this initial reconstruction could not be clearly observed any more. Thus it is not possible on these images to distinguish between the undisturbed sites and the surface defect sites for favorable InAs adsorption. On sample Ⓒ, at a surface coverage of 0.56 ML, domains of the underlying c(4×4) surface unit cells were only very rarely observed (an exemplary site ① is marked by a blue circle in Fig. 7.8). Though the signatures cover only about 50% of the surface area, the remaining surface appears deeply disturbed without any clear structure. The InAs signatures themselves, however, still appear very well aligned, clearly separated by periodic trenches that extend along the $[\bar{1}\,1\,0]$ direction. From a height profile ② in Fig. 7.8 the intermediate distance between these trenches is found to be about 1.2 nm in $[1\,1\,0]$ direction, corresponding to a threefold periodicity considering the (1×1) bulk-truncated surface unit cells. This apparently is incommensurate to the initial fourfold periodicity of the underlying c(4×4) reconstructed surface. Examining the typical alignment of the signatures in $[\bar{1}\,1\,0]$ direction, a fourfold symmetry still is favored (height profile ③ in Fig. 7.8).

7.3.3. Discussion: Formation and structure of the InAs signatures

As all signatures were found to be virtually of equal size, the average number of In atoms contained in one of these objects can be estimated. Considering the deposited In amount on each of the investigated samples with respect to the possible number of In atoms on the bulk-truncated InAs(0 0 1) surface and comparing this amount with the actual number of covered surface unit cells, the average In content of one signature must be about eight In atoms.

7.3. InAs thin films at submonolayer coverages

The InAs signatures are found at the hollow sites of the GaAs(0 0 1)-c(4×4) reconstruction as these sites are energetically more favorable for InAs adsorption than other sites at the reconstructed surface. This observation is supported by *density functional theory* (DFT) calculations of the potential-energy surface of GaAs(0 0 1)-c(4×4) considering a single diffusing In atom. The favorite initial adsorption site is marked by A_1 in Fig. 7.9 [180]. For the incorporation of more In atoms within one signature, a possible structural model will be discussed in the following.

Due to the missing As dimer, two group-III bulk lattice positions remain unoccupied on the bare GaAs-c(4×4) reconstructed surface at each hollow site. Moreover, the three characteristic As dimers are located closer to each other than in the respective bulk due to different dimer bond lengths to the As bulk atoms underneath [77] (see also Sect. 7.6.1). This provides two easily accessible incorporation sites for the — comparatively to Ga atoms — slightly larger In atoms. The locations of these sites are marked by ① in the illustrating structural model in Fig. 7.10. For the incorporation of further In atoms *ab initio* calculations [180] show that a splitting of the outer dimer back-bonds is a possible option. This would allow the incorporation of four additional In atoms per surface unit cell at positions marked by ② in Fig. 7.10.

Following this model, each hollow site at the undisturbed GaAs-c(4×4) reconstruction provides space for six indium atoms. Considering the bright contrast of the InAs signatures observed in the filled-state STM images and the As rich growth conditions it is very likely that additional As atoms are incorporated on top of the six In atoms to terminate some of their dangling bonds. Two possible configurations characterized by additional single As atoms or As dimers are illustrated at positions ③ and ④ in Fig. 7.10. Yet, as will be demonstrated in Eqs. 7.1, 7.2, and 7.3, neither configuration fulfills the ECR, since at least one excess electron is generated per signature.

- Configuration ①/② with solely six additional In atoms incorporated:

$$\begin{aligned}\text{electrons available}: &\quad 12 \cdot 3/4 \text{ e}^- \text{ (In}^{db}) + 6 \cdot 5/4 \text{ e}^- \text{ (As}^{db}) + 11 \cdot 2 \cdot 5/4 \text{ e}^- \text{ (As-As)} = 44 \text{ e}^- \\ \text{electrons required}: &\quad 12 \cdot 0 \text{ e}^- \text{ (In}^{db}) + 6 \cdot 2 \text{ e}^- \text{ (As}^{db}) + 11 \cdot 2 \text{ e}^- \text{ (As-As)} = 34 \text{ e}^- \end{aligned} \quad (7.1)$$

- Configuration ③ with additional single As atoms on top of the In atoms:

$$\begin{aligned}\text{electrons available}: &\quad 6 \cdot 3/4 \text{ e}^- \text{ (In}^{db}) + 12 \cdot 5/4 \text{ e}^- \text{ (As}^{db}) + 11 \cdot 2 \cdot 5/4 \text{ e}^- \text{ (As-As)} = 47 \text{ e}^- \\ \text{electrons required}: &\quad 6 \cdot 0 \text{ e}^- \text{ (In}^{db}) + 12 \cdot 2 \text{ e}^- \text{ (As}^{db}) + 11 \cdot 2 \text{ e}^- \text{ (As-As)} = 46 \text{ e}^- \end{aligned} \quad (7.2)$$

- Configuration ④ with additional As dimers on top of the In atoms:

$$\begin{aligned}\text{electrons available}: &\quad 4 \cdot 3/4 \text{ e}^- \text{ (In}^{db}) + 10 \cdot 5/4 \text{ e}^- \text{ (As}^{db}) + 13 \cdot 2 \cdot 5/4 \text{ e}^- \text{ (As-As)} = 48 \text{ e}^- \\ \text{electrons required}: &\quad 4 \cdot 0 \text{ e}^- \text{ (In}^{db}) + 10 \cdot 2 \text{ e}^- \text{ (As}^{db}) + 13 \cdot 2 \text{ e}^- \text{ (As-As)} = 46 \text{ e}^- \end{aligned} \quad (7.3)$$

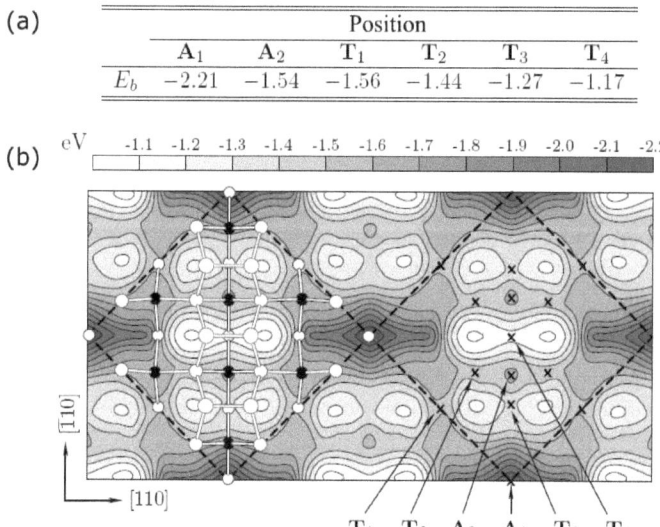

Figure 7.9: DFT calculations of the potential-energy surface for a single In adatom on GaAs(0 0 1)-c(4×4). (a) Calculated binding energy for selected surface sites at the (b) potential-energy landscape of the GaAs(0 0 1)-c(4×4) surface. The atomic positions are illustrated by an overlay of the structural model (Ga atoms: black, As atoms: white). The most favorable adsorption site is the center of the hollow site, marked by A_1. Image taken from Ref. [181].

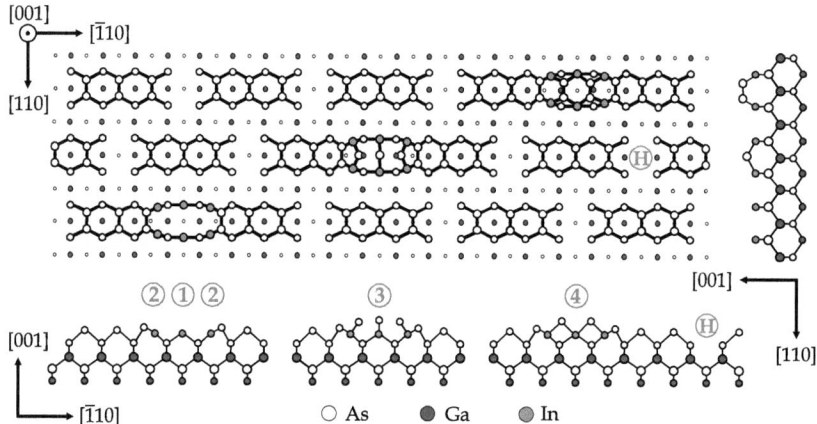

Figure 7.10: Top view and side views of structural models for the InAs signatures at the hollow site of the GaAs-c(4×4) surface. The center of the hollow site is marked by Ⓗ (identical to point A_1 in Fig. 7.9 b). Marks ① and ② indicate the most likely sites for the incorporation of In atoms. At positions ③ and ④ possible As terminations of dangling bonds are suggested.

These excess surface electrons could be responsible for the smeared electronic contrast observed in the STM images, which were all taken at negative sample bias (filled state images). Reversed tunneling conditions (empty state images) failed to provide an adequate STM signal at all, probably also due to these excess charges.

However, there could also be an alternative explanation for the origin of such effects. Considering additional adsorbed In atoms rather than an As termination this would lead to signatures with quasi-metallic electron states, resulting in the same consequences for the STM contrast. Moreover, these quasi-metallic In agglomerations could also host the averaged two more In atoms, which are missing in the suggested structural models above, assuming an averaged number of eight In atoms per observed signature.

On the other hand, at surface defect sites the number of adsorbed In atoms similar to ① and ② in Fig. 7.10 may increase due to shifts in the dimer alignment or additional missing top As dimers. This would explain the observed slight length extension in [$\bar{1}$ 1 0] direction of signatures bordering surface dislocations. Yet, due to the variability of such surface dislocations, completely different atomic arrangements of the In signatures seem also possible.

The present data eventually neither reveals the complete actual structure of the investigated InAs signatures nor suffices to prefer one of the proposed models. The signatures could not be resolved atomically, and the observed STM contrast from the filled state images supports either of the models.

Nevertheless, the observed disintegration of the underlying c(4×4) surface reconstruction and transformation of the local surface periodicity from fourfold to threefold in [1 1 0] direction at an InAs coverage of 0.56 ML may be supported by both argumentations. It is known that In atoms have a very high surface mobility comparable to the mobility of As atoms [182], thus In atoms are expected to move constantly from one adsorption site to the next during growth. These adsorption sites preferably are the c(4×4) hollow sites, where a number of In atoms very likely is already chemisorbed during the deposition process. When the sample is rapidly quenched, the remaining In quickly condenses at these hollow sites, where incorporated In atoms could act as attractors, but is not necessarily fully chemisorbed. The cumulated condensed In then partly remains quasi-metallic. With an increasing InAs amount deposited during growth, the number of populated adsorption sites rises and the mobility of the In atoms may be limited. The increasing density of the InAs signatures requires further surface space presumably at the cost of the As top dimers in between, a process which would lead to a quick dissolution of the the initial c(4×4) surface configuration. In result, there is no given symmetry pattern from the underlying initial surface any more that tends the remaining In atoms at the surface to condense in a fourfold periodicity pattern during quenching. Instead the transformation to another surface configuration might be energetically preferable during chemisorption, e.g. the threefold surface reconstruction observed on sample [C].

Indeed, with further InAs coverage the initial GaAs(0 0 1)-c(4×4) reconstructed surface will transform to — or be covered by — an InGaAs-(n×3) reconstructed surface, as will be shown in the following section of this work. However, such a possible transformation process is not understood in detail, yet.

In a previous STM study on the evolution of the InAs WL on GaAs(0 0 1)-c(4×4) bright signatures assembling at the hollow sites, comparable to the InAs signatures described in this work, were found and have been assigned to In atoms as well [178]. On the other hand, in the same report, a transformation of the GaAs(0 0 1)-c(4×4) surface unit cell to a (4×6) configuration was described occurring significantly at surface step edges, probably resulting from InGaAs alloying. Such observations could not be made in the investigations here, which may be due to differing growth conditions. The samples in Ref. [178] were grown at a comparatively low growth temperature and the presented images were taken at a 0.7° vicinal surface, which both is discussed to induce InGaAs alloying. Samples grown at a higher temperature in Ref. [178] (close to the $\beta 2(2\times 4)/c(4\times 4)$ transition point) show similar InAs signatures as observed here for an InAs coverage of 0.30 ML, yet the initial GaAs(0 0 1) exhibits a very unclear configuration. Unfortunately the atomic resolution of those STM data was not sufficient to support a conclusive discussion on the atomic structure of the surface and the observed signatures. The proposed structural model for In incorporation in Ref. [178] is thus solely based on the low temperature data and is not in agreement with the experimental findings here.

7.3.4. Summary

The deposition of very low amounts of InAs, e.g. from 0.09 ML to 0.56 ML, onto the GaAs(0 0 1)-c(4×4) reconstructed surface leads to the formation of characteristic signatures of InAs, that each contain about eight In atoms in average. These signatures are adsorbed at the hollow sites of the c(4×4) reconstruction, but preferentially assemble at surface defect sites or domain boundaries. With increasing amounts of InAs the number of signatures increases proportionally, and the signatures progressively occupy the hollow sites of the undisturbed GaAs-c(4×4) surface. A model for the initial formation of these signatures is introduced by discussing the incorporation of the first six In atoms at the hollow site with the help of DFT and *ab initio* calculations. Models of the final signature structure are discussed, but could not be clarified from the vague STM data, which supports either of the discussed models. The vagueness of the STM data at the signature sites may be due to electronic effects from excess electrons or the possible quasi-metallic character of the signatures themselves.

7.4. InAs thin films at coverages close to one monolayer

7.4.1. Sample growth parameters

During growth of the following sample $\boxed{D}$, after a deposition time that corresponds to roughly about 0.7 ML of deposited InAs, a transformation of the RHEED pattern was observed for reflexes in both surface directions. The observed pattern clearly distinguished from the initial c(4×4) related pattern indicating a significant change of the global surface reconstruction. However, the rather smeared contrast of the reflexes did not allow a more detailed analysis of this transformation or the resulting pattern.

The InAs growth was interrupted shortly afterwards this observation. The respective growth parameters are given in Tab. 7.2.

7.4.2. STM results

Overview

Figure 7.11 a shows a large scale STM image of the initial GaAs(0 0 1)-c(4×4) surface after the deposition of about 0.73 ML InAs. The surface appears clean and rather smooth, exhibiting larger plateaus and neighboring flat terraces separated by steps of monoatomic height. On top of the terraces small spots of bright contrast appear distributing throughout the observed surface areas.

Figure 7.11 b presents a magnified image of the surface area, with the respective location on the large scale image (a) marked by a blue square. There are two phenomena clearly observed in this image.

Firstly, there is an underlying surface of a configuration of brick-like objects that can be clearly distinguished from the initial GaAs-c(4×4) reconstruction. This configuration appears very stable and well ordered throughout all of the observed terrains. As indicated by the RHEED observations it is evident that the surface has fully transformed during growth. Thus the resulting surface suggests that the deposited InAs has induced the formation of a first complete atomic layer on top of the initial surface. This first layer actually is formed by an InGaAs alloy, as will be shown later in the discussion.

Secondly, there are small objects of bright contrast, quite similar to the InAs signatures observed at lower InAs coverages. Hence it is consequent to assume that these small objects originate from InAs as well. The objects cover about 15 % of the surface area which in consensus with the previously formed first InGaAs atomic layer then corresponds to a total coverage of 115 % of the initial GaAs sample surface.

		sample D
deposition time τ		95 s
calculated InAs deposition derived from the growth function		0.73±0.06 ML
1st layer	coverage by (4×3) unit cells *	100% (by def.)
	estimated InAs content	0.67 ML (by def.)
2nd layer	coverage by (2×4) unit cells **	15±5%
	estimated InAs content	0.11±0.04 ML
estimated total InAs content derived from the STM data		0.78±0.04 ML

Table 7.2: *InAs growth parameters for sample D in the investigation stage of InAs coverages above 0.67 ML. The presented data relies on the findings in this work and follows the discussion in Appendix B.*
*The surface area of one (4×3) surface unit cell corresponds to $A = 1.92$ nm^2.
**The surface area of one (2×4) surface unit cell corresponds to $A = 1.28$ nm^2.

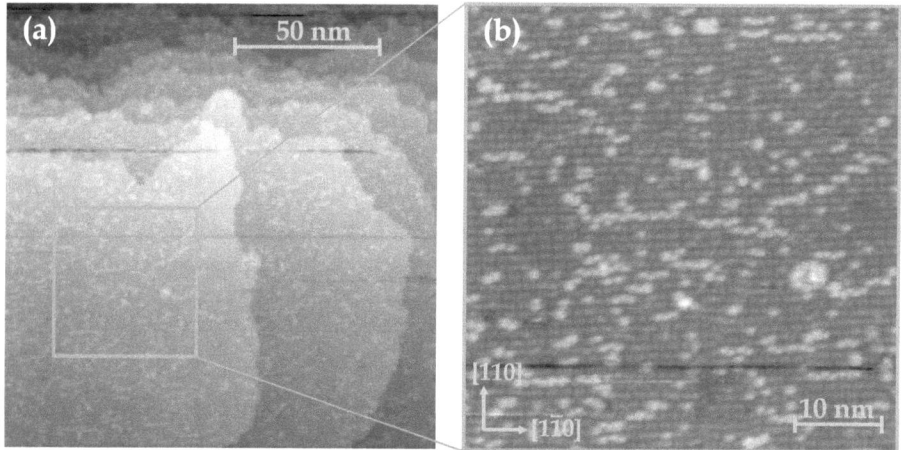

Figure 7.11: *Sample D: Filled state STM images ($V_T = -2.7$ V, $I_T = 0.2$ nA) of a 0.73 ML InAs coverage on the GaAs(0 0 1)-c(4×4) surface. (a) On a large scale image a smooth surface with flat terraces of monoatomic height is observed, exhibiting small signatures of bright contrast. An atomically resolved magnification of a surface area (marked by the blue square) is depicted in (b) showing that the surface has clearly changed exhibiting a new surface configuration.*

The InAs objects appear well aligned in respect to the underlying surface reconstruction and distribute evenly, comparable to very small two-dimensional islands. It appears as if they initialize the growth of a second InAs atomic layer.

Structural details of the first InGaAs atomic layer

The STM image in Fig. 7.12 shows the surface reconstruction of the InGaAs ML in detail with atomic resolution. As in filled state images mostly As contributes to the observed

7.4. InAs thin films at coverages close to one monolayer

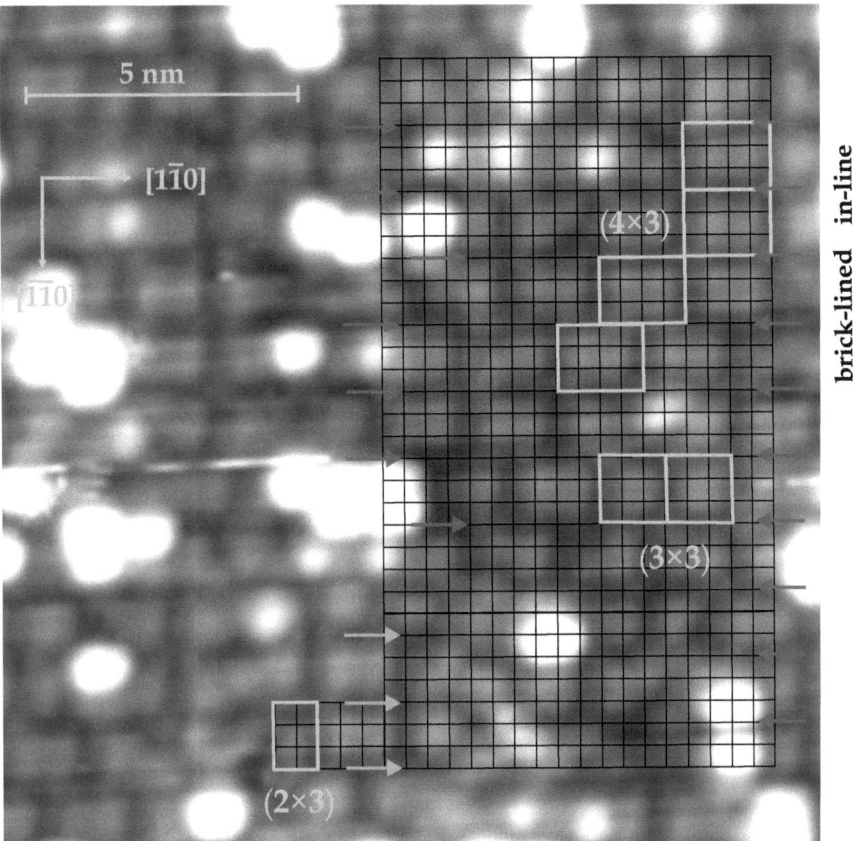

Figure 7.12: Sample ⟦D⟧: Filled state STM image ($V_T = -2.7\,V$, $I_T = 0.2\,nA$) of the ($n \times 3$) reconstructed InGaAs ML. The auxiliary lattice partly depicted in the image illustrates surface unit cells of the bulk-truncated GaAs(0 0 1) surface. The different types of apparent ($n \times 3$) surface unit cells and their possible alignment are exemplarily marked by yellow boxes. The periodic position of the significant surface trenches is marked by magenta arrows. A shift in the periodicity of the trench location is illustrated by blue arrows.

STM contrast, the brick-like objects observed here are probably quite similar to the As dimer bricks known from the GaAs-c(4×4) surface reconstruction and thus correspond to As dimer alignments as well.

Along the [1 1 0] direction the periodicity of surface unit cells is characterized by trenches clearly separating the single As dimer bricks, remaining remarkably undisturbed throughout the bare surface. The positions of these trenches are marked by magenta arrows. Applying an auxiliary lattice representing the bulk-truncated surface unit cells of the GaAs(0 0 1) surface (surface lattice parameter $a = 0.40\,nm$) reveals that the distance between these trenches corresponds to three bulk-truncated surface unit cells, leading to an (n×3) periodicity of the InGaAs ML surface in respect to the [1 1 0] direction.

Occasionally disturbances in this trench alignment were observed, typically near the adsorbed InAs islands on top of the InGaAs ML. In the lower part of the STM image in Fig. 7.12 blue arrows indicate such a shift in the trench position.

In [$\bar{1}$ 1 0] direction the periodic alignment of the As dimer bricks is not as undisturbed. The surface unit cells are clearly separated from each other, probably due to a missing As dimer. Yet the position of this missing dimer often locally varies throughout the surface. The typical configurations of unit cells that were found are illustrated by yellow boxes in Fig. 7.12. The mostly occurring configuration is an in-line alignment of (4×3) unit cells, but also a brick-lined alignment was observed, both extending by four bulk-truncated surface unit cells in [$\bar{1}$ 1 0] direction. Pairwise (3×3) unit cells were found occasionally, allowing the observed alignment shift of the (4×3) unit cells between in-line and brick-lined. Rarely, also (2×3) unit cells were observed.

To reveal more information on this InGaAs-(4×3) surface, local images with high atomic resolution were taken. In Fig. 7.13 a a typical area of the covering InGaAs surface is presented in more detail. The size and location of the typical surface unit cell configurations, e.g. (4×3) and (3×3), are indicated by boxes of dotted yellow lines.

From the (4×3) periodicity it is assumed that the surface is characterized by a parallel alignment of three As dimers and a dimer vacancy site in [$\bar{1}$ 1 0] direction, very similar to the As dimer alignment of the GaAs-c(4×4) reconstruction. However, the top dimers observed here show a slightly darker contrast at the position of the central dimer. Occasionally a local depression of darker contrast is found on one of the outer top dimers (as marked by a yellow arrow in Fig. 7.13 a), most likely caused by a *heterodimer* (hd).

In [1 1 0] direction it is assumed that the triple As dimer blocks alternate with a single atomic trench. Yet, within the trench weak protrusion can be observed, slightly darker than the contrast of the dimer bricks. This contrast probably originates from another As dimer with an orientation rotated by 90° as compared to the top dimers, that is located on a lower surface level and thus appearing less bright.

Height profiles were taken along the [$\bar{1}$ 1 0] direction at the top dimer bricks ① as well as at the trench dimers ②. The corresponding data is presented in Fig. 7.13 b.

For the As top dimer bricks ① a distance of about 1.6 nm between dimer vacancies corresponding to four bulk-terminated surface unit cells can be derived from the data. The height difference between the top dimers and the trench dimers is about 0.06 nm. Furthermore, the height contours also show the slightly lower position of the central top dimer as compared with both outer dimers with a height difference of about 0.01 nm.

For the As trench dimers ② a clear correlation between the top As dimer structure and the trench dimers is visible in the height profile. Each set of a triple dimer block and a dimer vacancy site are accompanied by two trench dimers. Following this, the rarely observed (2×3) unit cell results if the central top dimer is missing, and thus the (2×3) surface unit

7.4. InAs thin films at coverages close to one monolayer

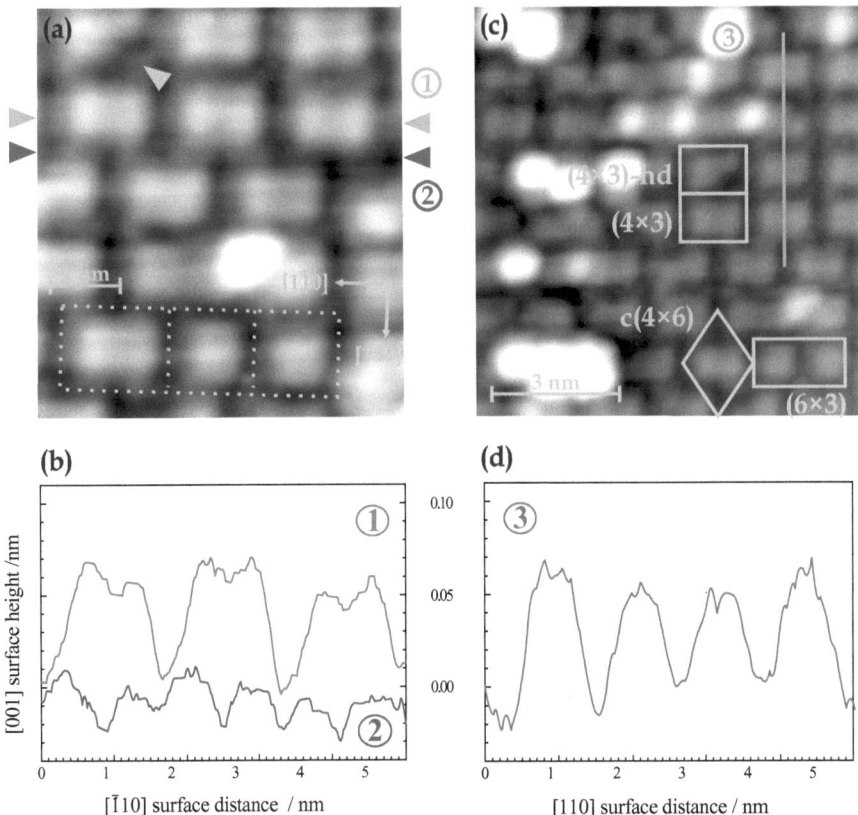

Figure 7.13: *Sample* D: *(a) and (c) Filled state STM images ($V_T = -2.7\,V$, $I_T = 0.2\,nA$) of the (4×3) reconstructed InGaAs ML in more detail. In image (a) the contrast of the assumed triple As top dimers is clearly observed. The yellow boxes exemplarily depict the surface unit cells. A yellow mark points to a contrast depression presumably caused by a heterodimer. Height profiles were taken along the [1̄10] direction on the As triple dimers (between the blue marks ①) and in the surface trench (between the magenta marks ②). The respective data is shown in (b). In image (c) the different arrangement and resulting periodicity of surface unit cells is displayed by yellow boxes. The data from a height profile taken perpendicular to the trenches ③ is shown in (d).*

cell is characterized by only one top dimer and one trench dimer. A (3×3) surface unit cell, on the other hand, then is characterized by two top dimers, but has to share one of two trench dimers due to the twofold periodicity of the trench dimers. This asymmetry requires a pairwise appearance of (3×3) unit cells, which consequently have to be referred to as one (6×3) surface unit cell.

The local insertion of a (6×3) or the more rarely observed (2×3) surface unit cell leads to a shift in the periodic alignment of the triple As dimer blocks by two bulk-terminated surface unit cells in [1̄10] direction. This leads to a shift in the surface periodicity from the

(4×3) in-line to a brick-lined configuration, which consistently can be described by a c(4×6) surface unit cell. Such a configuration can be observed in the STM image in Fig. 7.13 c where the respective alignment of surface unit cells is illustrated by yellow boxes. An appearing (4×3)-hd configuration is marked as well.

From a height profile in [1 1 0] direction ③ the lateral distance of the (n×3) periodic dimer trenches can be derived. The corresponding data is shown in Fig. 7.13 d. The derived average distance is about 1.2 nm corresponding to three bulk-terminated surface unit cells.

Structural details of the small InAs islands on top of the InGaAs monolayer

The focus of the STM images in Fig. 7.14 is on the small InAs related islands that are observed on top of the InGaAs-(4×3) reconstructed ML. It will be later shown, that these islands indeed contain pure InAs. The larger scale image (Fig. 7.14 a) demonstrates the clear preference of these islands to align in rows along the [1 $\bar{1}$ 0] direction. Comparable rows in [1 1 0] direction could not be observed.

In Fig. 7.14 b a magnified section of the surface is shown. It appears that there are basically two types of InAs objects with respect to their observed size. There is a smaller type, which is mostly observed. When aligned it almost exclusively forms zig-zag chains in [1 $\bar{1}$ 0] direction, as marked in blue by ①. The other type appears larger, about double the size of the smaller type. At a very close look it appears, as if it consists of two of the smaller InAs objects located very close to each other in [1 1 0] direction. These larger InAs objects also align along the [1 $\bar{1}$ 0] direction, yet not in zig-zag alignment and with much shorter chains.

The otherwise very stable (n×3) periodicity of the trench of the underlying InGaAs-(4×3) surface is found to be disturbed underneath significant accumulations of InAs signatures. This is illustrated in Fig. 7.14 b by the arrangement of the dotted yellow lines depicting the trench positions.

Comparing the size of the (4×3) surface unit cell with a typical zig-zag chain alignment of InAs objects, as marked by the yellow box in Fig. 7.14 b, indicates their periodicity along the [1 $\bar{1}$ 0] direction. A pair of two objects perfectly fits the boundaries of the (4×3) surface unit cell in [1 $\bar{1}$ 0] direction, thus a periodic alignment of (2×n) can be derived for the InAs objects. However, as there is no continuous alignment of InAs objects observed in [1 1 0] direction, e.g. chains, a conclusion on the n-fold periodicity cannot be drawn for this direction yet.

There are previous reports of STM studies on InGaAs MLs, which show some similarities with the experimental results in this work considering specific structural aspects [106, 107, 183]. In general these reports present findings of In-rich (2×4) reconstructed domains, preferentially aligning in chains upon an — assumed — (4×3) reconstructed InGaAs surface alloy. Yet, there are significant differences concerning the sample growth. The related findings are solely based on lattice-matched InGaAs alloys grown on InP(0 0 1) [106, 183] and

7.4. InAs thin films at coverages close to one monolayer

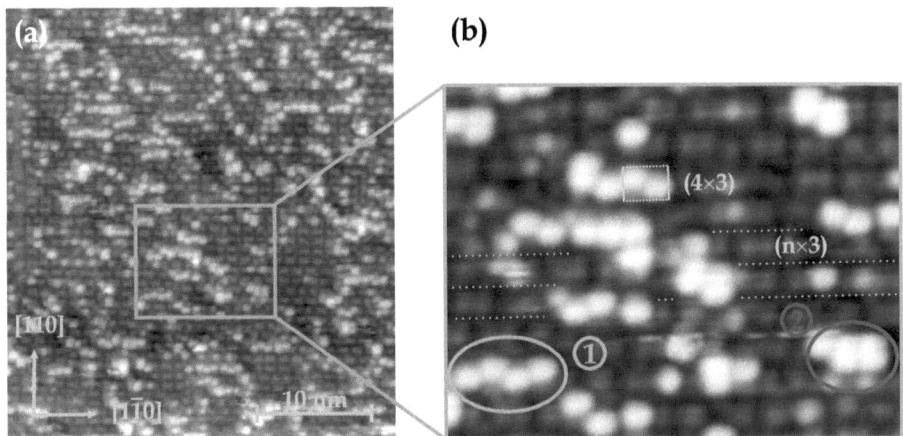

Figure 7.14: Sample ⓓ: Filled state STM image ($V_T = -2.7\,V$, $I_T = 0.2\,nA$) of the small InAs islands that form on the (4×3) reconstructed InGaAs ML. (a) The larger scale image shows the typical distribution of these InAs islands covering about 15% of the surface area. They preferentially align in chains along the [$\bar{1}$ 1 0] direction. A typical surface area, as marked by the green box, is magnified in (b). It shows the two basic types of such objects: ① a smaller type arranging in zig-zag chains and ② a larger type of about double the size of the small islands usually aligned in-line. The two larger type islands in ②, neighboring along the [$\bar{1}$ 1 0] direction, can be clearly separated in by a slight depression in STM contrast in between. However, such a slight contrast depression can be also observed in the perpendicular direction, leading to a cloverleaf-like appearance of the whole structure. According to these contrast observations it might be presumed that each of the larger islands actually consists of two of the smaller islands. The (4×3) surface unit cell of the underlying InGaAs surface is illustrated by a yellow box. Dotted yellow lines depict the locations of the associated (n×3) periodic trench.

on GaAs(0 0 1) [106, 107] with a deposited material amount of 25 ML InGaAs. Furthermore, unusual growth conditions were chosen, such as a V/III ratio of only 2.7 for the growth on GaAs(0 0 1) despite the use of an As$_4$ source [106].

The presented atomic resolution was only suboptimal and not sufficient to reveal the atomic structure of the (4×3) reconstructed InGaAs surface or hence conclusively discuss a compatible structural model. However, the studies of InGaAs on the GaAs substrate in Refs. [106, 107] further report on $\alpha2(2\times4)$ reconstructed unit cells aligning in zig-zag chains that were observed. The proposed structure in these reports was derived from a well known configuration on the bare InAs(0 0 1) surface [184]. These reported findings seem very similar to the observations on the small InAs islands in this work, which likewise align in zig-zag chains (e.g. Fig. 7.14).

7.4.3. Discussion: Formation and structure of the $In_{2/3}Ga_{1/3}As$ monolayer

There is a number of proposed structural models for an (n×3) reconstructed InAs or InGaAs surface, some of which are discussed in a DFT study on ultra thin InGaAs films on GaAs(0 0 1) with respect to the chemical potential of the environmental As during growth as well as the mechanical strain caused by the atomic configuration of that film [185]. The discussed models all refer to an $In_{2/3}Ga_{1/3}As$ alloyed ML. Yet among those there is only one that fits the STM observations presented in this work sufficiently. Its detailed atomic structure is shown in Fig. 7.15.

Basically in this model, the last bulk-truncated group-III layer consists of a composition of two thirds of In atoms and one third of Ga atoms terminated by a complete As layer, and due to the As rich conditions additional As dimers are formed on top [186]. The group-III layer is characterized by a strict alignment of two rows of In atoms (depicted in orange) along the [$\bar{1}$ 1 0] direction alternating with one row of Ga atoms (depicted in blue). Atomic interchange processes during growth were reported to prefer such a very strict two rows In atoms/one row Ga atoms configuration within the reconstructed $In_{2/3}Ga_{1/3}As$ ML [187]. An arrangement like this therewith defines the (n×3) periodicity in the [1 1 0] direction.

The following complete As layer arranges at the designated bulk positions above the row of Ga atoms, with each As atom back-bonded to one Ga atom and one In atom. In the positions centering above the two In atom rows, however, As dimers oriented in [$\bar{1}$ 1 0] direction are formed with each As dimer atom back-bonded to two In atoms. The resulting longitudinal row of dimers is left uncovered thus forming a trench, corresponding to the trench that was observed in the STM studies, while the rest of the surface is covered by additional As dimers.

These As dimers are oriented along the [1 1 0] direction. They form groups of three dimers followed by a dimer vacancy site (hollow site) arranged along the [$\bar{1}$ 1 0] direction. In this way the triple dimers establish a (4×n) periodic chain, which is located vertically directly above the rows of Ga atoms.

Within these chains, the As triple dimers can be arranged in two ways in relation to the neighboring chains; in either a parallel in-line alignment corresponding to a (4×3) reconstruction or a shifted brick-lined alignment corresponding to a c(4×6) reconstruction. In either way the shift along the [$\bar{1}$ 1 0] direction must be an even multiple of the bulk-truncated surface unit cell to be compatible with the (2×n) periodicity of the trench dimers. Interestingly, shifts by an odd multiple (e.g. one or three bulk unit cells) were not observed in the studies here, which indicates a high stability of the basically (2×3) reconstructed trench dimers.

The occasionally observed (6×3) surface unit cells can easily be generated by removing the third top dimer, which results in an apparent (3×3) surface unit cell. As this would be incommensurate to the trench dimer periodicity, a neighboring second (3×3) surface unit

7.4. InAs thin films at coverages close to one monolayer

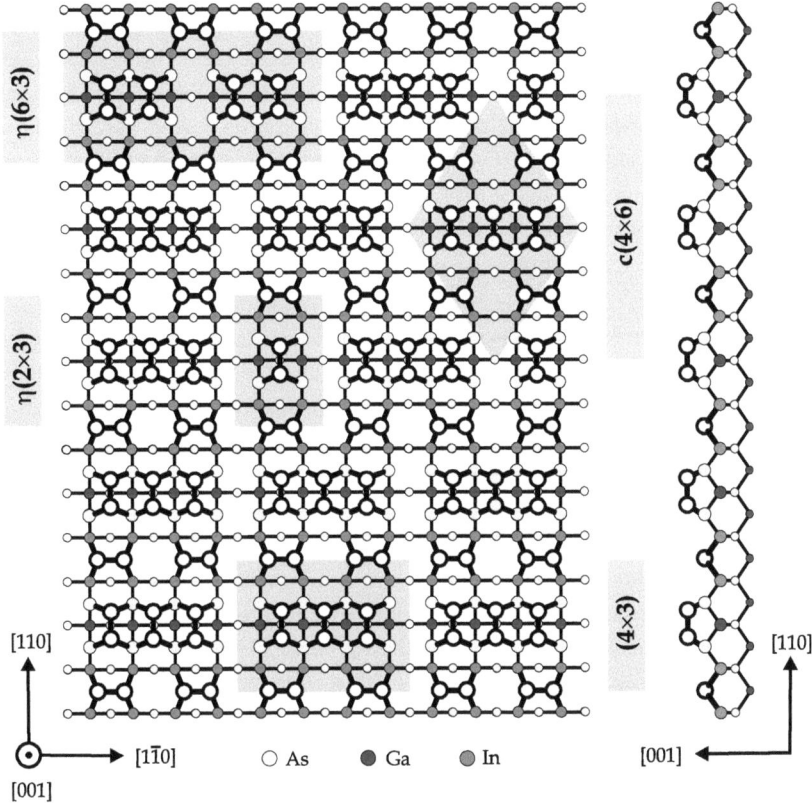

Figure 7.15: Top view and side views of the proposed structural model of the (4×3) reconstructed In$_{2/3}$Ga$_{1/3}$As ML. Atoms below the figure plane are depicted by smaller circles. The different observed surface unit cells are marked by yellow boxes. In the lower part of the model, the typical in-line (4×3) arrangement is shown. In the center of the illustrated model an η(2×3) surface unit cell leads to an alignment shift to the brick-lined arrangement described by a c(4×6) unit cell. In the upper part of the model the η(6×3) unit cell is shown, leading to the same alignment shift. The denotation η is introduced to distinguish these particular structures from structures suggested elsewhere (e.g. in Ref. [104]).

cell is generated likewise, both sharing one of the trench dimers. In the same way, the rarely observed (2×3) surface unit cell then exhibits only one top As dimer, yet is compatible with the trench dimer alignment. Both of these structural anomalies were found to be responsible for a transition between the two alignments of the (4×3) surface unit cells.

In Fig. 7.16 this structural model is compared with the contrast of typical STM images deducted from this work. The accordance is evident, including the contrast from the trench dimers. The different surface unit cells are marked by dotted colored boxes.

As further stated in the experimental findings, the height difference between the triple As top dimer block and the trench dimer was measured as 0.06 nm. This is slightly less that the calculated value of 0.085 nm from DFT calculations [185] or the experimental value of 0.075 nm from XRD experiments [105]. This difference can be related to electronic effects and/or the apex curvature of the STM tip leading to additional electron tunneling from the neighboring As top dimers into the flanks of the tip apex when imaging above the trench. Thus the trench dimer might appear less deep than it actually is.

With respect to previous reports, related aspects of the present findings shall be discussed in the following.

Considering the surface formation energy of an $In_{2/3}Ga_{1/3}As(0\,0\,1)$ film on the $GaAs(0\,0\,1)$ surface, the DFT calculations in Ref. [185] slightly prefer the (2×3) reconstruction, as suggested by Ref. [104], over the (4×3) reconstructed structural model suggested here. However, considering the typical accuracy of DFT calculations, the small energy difference between both surface structures may diminish.

This originally assumed (2×3) structural model from Ref. [104] is not in full agreement with the proposed structure for the (2×3) surface unit cells here. Whereas the model in Ref. [104] assumes two top dimers and one trench dimer, it was found in this work that the (2×3) structure rather derives from the (4×3) structure and is characterized by only one As top dimer and one As trench timer. To allow a better distinction of the different models the denotation $In_{2/3}Ga_{1/3}As(0\,0\,1)$-$\eta$(2×3) is thus introduced. Likewise the denotation $In_{2/3}Ga_{1/3}As(0\,0\,1)$-$\eta$(6×3) is introduced for the assumed structure of the (6×3) unit cell, as there are different atomic configurations thinkable. This might for instance be an In–In top dimer occupying the central hollow site, which would not be clearly distinguishable in filled-state STM images. Yet, such an ι(6×3) configuration [188] seems unlikely considering the highly As rich growth conditions.

Neither of the three surface reconstructions presented here, i.e. the (4×3)/c(4×6), the η(6×3), nor the η(2×3) reconstruction, fulfill the ECR (Table 7.3). All of them are characterized by a deficit in electrons. However, there are reports on stable GaAs or InAs surface reconstructions that likewise violate the ECR by missing electrons [39, 60]. It might be added that the previously assumed (n×3) structural models also do not fulfill the ECR, but are characterized by excess electrons instead [188].

Considering the ECR yet another experimental finding may be explained. Both the η(2×3) and the η(6×3) configuration are found to induce the shift of the (4×3) surface unit cells between the in-line and brick-lined arrangement. Nevertheless, the η(6×3) configuration is much more often observed. Comparing the results of the ECR per (2×3) unit cell surface section, it becomes clear that the η(2×3) reconstruction has the highest electron deficit ($-3\,e^-$ per section) among the three configurations. The electron deficit of the η(6×3) reconstruction is much less ($-1\frac{2}{3}\,e^-$ per section) which seems to be energetically favorable. The (4×3)/c(4×6) reconstruction itself has the least electron deficit ($-1\,e^-$ per section) and is probably therefore dominating the surface structure.

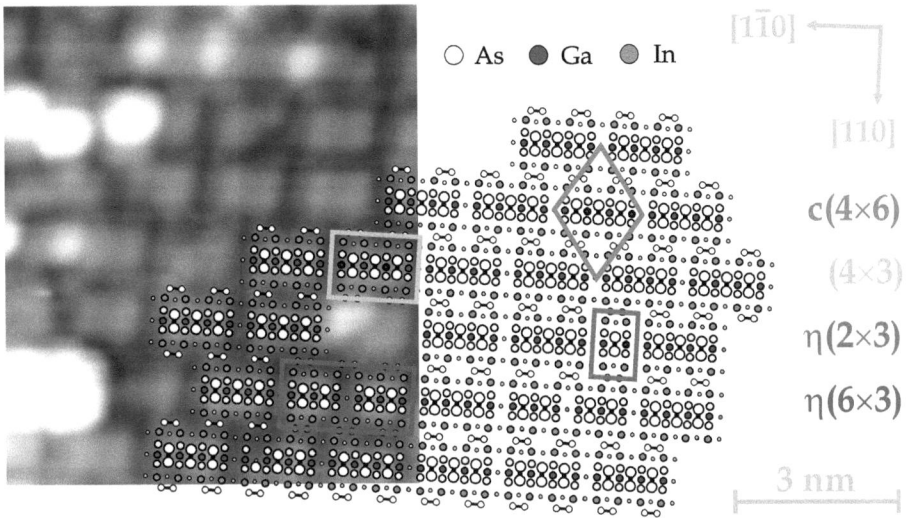

Figure 7.16: Sample ⓓ: Filled state STM images ($V_T = -2.7\,V$, $I_T = 0.2\,nA$) of the (4×3) reconstructed $In_{2/3}Ga_{1/3}As$ ML overlaid by the discussed structural model. Atoms below the figure plane are depicted by smaller circles. The different types of surface unit cells are illustrated by colored boxes.

	electron balance at respective structural building blocks		
	one As top dimer	one As trench dimer	one dimer vacancy
electrons available	$5 \cdot 2 \cdot 5/4\,e^-$ (As–As) $2 \cdot 5/4\,e^-$ (As^{db}) $= 15\,e^-$	$1 \cdot 2 \cdot 5/4\,e^-$ (As–As) $2 \cdot 5/4\,e^-$ (As^{db}) $= 5\,e^-$	$4 \cdot 5/4\,e^-$ (As^{db}) $= 5\,e^-$
electrons required	$5 \cdot 2\,e^-$ (As–As) $2 \cdot 2\,e^-$ (As^{db}) $= 14\,e^-$	$1 \cdot 2\,e^-$ (As–As) $2 \cdot 2\,e^-$ (As^{db}) $= 6\,e^-$	$4 \cdot 2\,e^-$ (As^{db}) $= 8\,e^-$
electron balance	$+1\,e^-$	$-1\,e^-$	$-3\,e^-$

	electron balance of selected surface unit cell configurations				
unit cell configuration	top dimers	trench dimers	dimer vacancies	\multicolumn{2}{c}{electron balance}
				total	per (2×3) section
(4×3)/c(4×6)	$3 \cdot (+1\,e^-)$	$2 \cdot (-1\,e^-)$	$1 \cdot (-3\,e^-)$	$-2\,e^-$	$-1\,e^-$
$\eta(6\times3)$	$4 \cdot (+1\,e^-)$	$3 \cdot (-1\,e^-)$	$2 \cdot (-3\,e^-)$	$-5\,e^-$	$-1\tfrac{2}{3}\,e^-$
$\eta(2\times3)$	$1 \cdot (+1\,e^-)$	$1 \cdot (-1\,e^-)$	$1 \cdot (-3\,e^-)$		$-3\,e^-$

Table 7.3: ECR balance of selected surface configurations corresponding to the structural alignments of the surface unit cells on the $In_{2/3}Ga_{1/3}As(0\,0\,1)$ surface observed in the STM data. In the upper table calculations apply to common structural building blocks of the different reconstructions. In the lower table the ECR balances of the different reconstructions are given and additionally calculated in reference to a (2×3) unit cell surface section.

Diffraction experiments mostly report on a (1×3) periodicity in RHEED [71,103] or a (2×3) periodicity in grazing incidence XRD [104,105] investigations of the InGaAs ML. Due to the structural properties and the technique for deriving diffraction patterns, this has not to be in conflict with the findings derived here. In [1 1 0] direction the observed trench is very stable throughout the surface and its long range (n×3) periodicity thus nicely contributes to the diffraction patterns. In the perpendicular direction along [$\bar{1}$ 1 0], the frequent shift of the (4×3) unit cell between in-line and brick-line accompanied with the insertion of $\eta(6\times3)$ and $\eta(2\times3)$ unit cells for compensation significantly disturbs the long range periodicity. Diffraction methods like grazing incidence XRD average over a certain surface area, and if neighboring (4×3) surface unit cells are locally shifted, only the common lattice of trench dimers in both alignments contribute to the diffraction intensity resulting in a (2×3) pattern. In RHEED the penetration depth is much shorter than in XRD, and so the As top dimers may practically shade the trench dimers from the incident electron beam perpendicular to the trench direction. Thus the twofold periodicity of the trench dimers along the [$\bar{1}$ 1 0] direction cannot sufficiently contribute to the diffraction intensity and a (1×3) pattern may be observed.

Finally it should be added that the formation of an $In_{2/3}Ga_{1/3}As$ alloy is assumed even though only pure InAs was deposited during growth. The additionally needed Ga atoms thus might originate from the bulk lattice by dissolving material at the surface step edges and consuming the terraces during growth. Such a so-called step flow growth was well observed in the InAs/GaAs system before [182].

7.4.4. Discussion: Formation and structure of the InAs islands on top of the $In_{2/3}Ga_{1/3}As$ monolayer

The proposed structural models in Refs. [106,107,183], investigating InAs coverages on comparable surfaces, have proven very consistent with the experimental findings in this work. For the significant zig-zag chains an $\alpha2(2\times4)$ reconstruction is assumed. Its well established detailed structure is shown in Fig. 7.17 a [184,189,190].

The $\alpha2(2\times4)$ reconstructed surface unit cell evolves on top of the As terminated last layer of the bulk lattice. Six In atoms constitute the following group-III layer, which is not fully covering the surface area. At the bulk-designated positions of the vacant seventh and eighth In atoms a trench is formed exhibiting a characteristic As dimer within.

This As trench dimer is oriented in the [$\bar{1}$ 1 0] direction and very similar to the trench dimer on the (4×3) surface discussed before. In fact it is most likely identical to that trench dimer, backbonded to four In atoms, as the alignment of the InAs signatures in the STM image in Fig. 7.18 clearly adapts to the underlying (4×3) reconstructed layer.

On top of the In atom row located at the outer side of the surface unit cell towards the trench, one As dimer is formed, bonding to four In atoms. This As dimer is oriented along the [$\bar{1}$ 1 0] direction similar to the trench dimer. The remaining two In atoms then form two In dimers with their respective In neighbors in [1 1 0] direction.

The observed image contrast of such a structure in filled-state images is mostly dominated by the As top dimer. Thus it is evident that a small modification of this model laterally moving the top dimer onto the inner In atoms of the surface unit cell, further denoted as $\alpha 2(2\times 4)$-m, can perfectly account for the observed zig-zag chains, by alternating both of these structures (Fig. 7.17b). Consequently, the observed larger InAs signatures then should correspond to an As double dimer on top of the In atoms. This well-known $\beta 2(2\times 4)$ structure is illustrated in Fig. 7.17c.

There are corresponding DFT calculations that provide the expected STM contrast for filled-state images of both the InAs(0 0 1)-$\alpha 2(2\times 4)$ and $\beta 2(2\times 4)$ surface reconstructions [184]. In Fig. 7.18 the calculated STM contrast of the proposed structures is compared with the contrast of typical STM images deducted from this work. It shows that both the STM contrast and the DFT calculation agree nicely.

It may be assumed from the observed STM contrast that the initial alignment of the (2×4) reconstructed unit cells is determined by the trench dimers of the underlying (4×3) reconstruction. In this model the growth starts by replacing the triple As top dimers by In atoms, while the adjacent trench row is left uncovered. In the [$\bar{1}$ 1 0] direction the top (2×4) and the underlying (4×3) have an even periodicity and thus the structures are well commensurate. However this is not the case in the [1 1 0] direction due to the respective threefold and fourfold periodicities. If one surface trench is left uncovered the neighboring

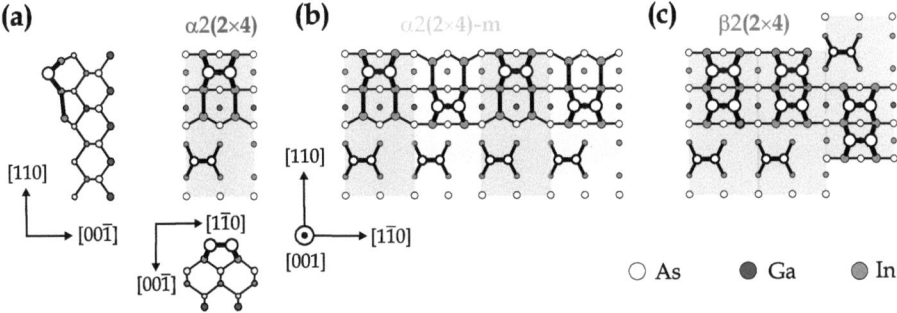

Figure 7.17: *Structural models for the InAs-(2×4) reconstructed islands following Ref. [189]. (a) Top view and side views of the InAs $\alpha 2(2\times 4)$ surface unit cell. (b) Alignment of the $\alpha 2/\alpha 2$-m configuration as attributed to the observed zig-zag chains. (c) In-line alignment of $\beta 2(2\times 4)$ surface unit cells and an occasionally observed mirror alignment corresponding to the threefold symmetry of the trench dimer of the underlying $In_{2/3}Ga_{1/3}As$ ML. The characteristics of the underlying $In_{2/3}Ga_{1/3}As$ (4×3) is shown to illustrate the specific alignment of the InAs islands on top of that first ML. Atoms below the figure plane are depicted by smaller circles.*

trench then must be covered by In atoms to fit to the (2×4) periodicity. The In atoms are the immediately covered by As dimers in either $\alpha2$ or $\alpha2$-m arrangement and form the typically observed zig-zag chains.

The occasionally observed $\beta2$ arrangement is most likely due to the highly As-rich conditions during growth. However, the $\beta2$-(2×4) reconstruction was not observed in Refs. [106, 107] for growth on GaAs. This may be due to the reported V/III ratio of as little as 2.7, i.e. the comparatively As-poor conditions.

The STM studies occasionally showed structural shifts by one bulk surface unit cell in the otherwise very stable threefold periodicity of the trench on the (4×3) reconstructed surface

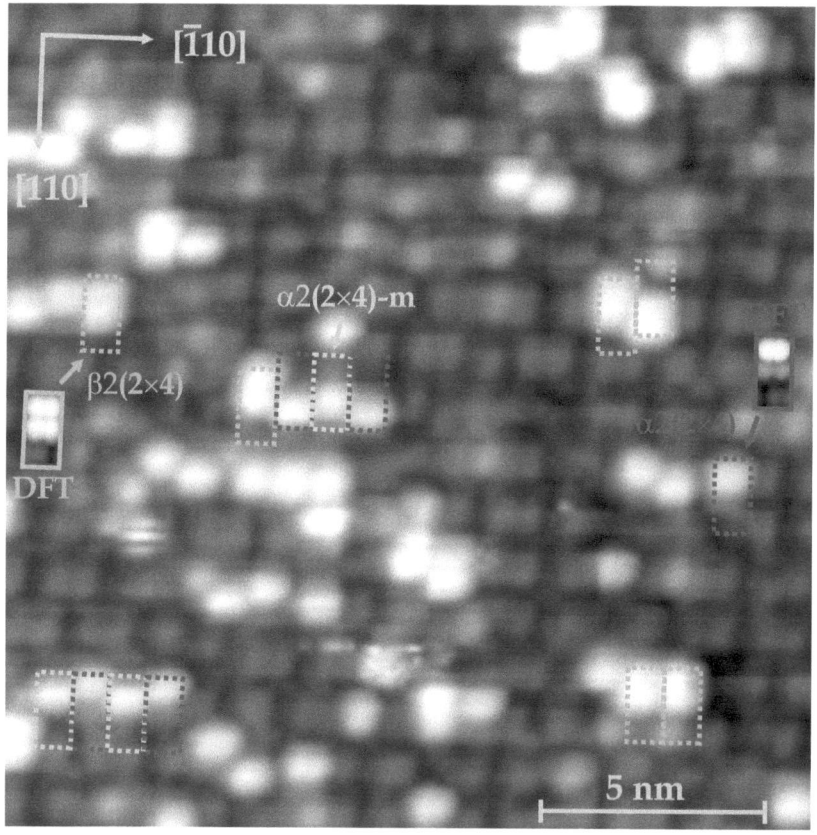

Figure 7.18: sample $\boxed{D}$: Filled state STM image ($V_T = -2.7\,V$, $I_T = 0.2\,nA$) of the $In_{2/3}Ga_{1/3}As$-(4×3) reconstructed ML with additional InAs-(2×4) reconstructed islands on top, corresponding to a total coverage of about 0.73 ML. The alignment of the three different island structures — the $\alpha2$, $\alpha2$-m, and $\beta2$ configurations — are exemplarily marked by dotted colored boxes according to the color scheme used in Fig. 7.17. The expected STM contrast derived from DFT calculations [184] is shown in the straight boxes of the respective color.

and comparable local surface dislocations affecting the trench position (e.g. Fig. 7.14b). Most of this disorder in the (n×3) periodicity seems to occur very close to or underneath areas covered by the (2×4) unit cells. However, assuming effects of the adsorbed (2×4) surface unit cells onto the underlying (4×3) reconstructed surface to shift the position of the trench seems unlikely, as this requires a further restructuring of large adjacent areas, to obtain the stable threefold trench symmetry that was observed otherwise. It seems more likely that the underlying surface is not as stable as suspected and that shifts in the (n×3) periodic arrangement of the trenches do occur. Such a shift of the trench location in [1 1 0] direction then generates a local fourfold periodicity at the local defect site. This might induce a preferable adsorption of the (2×4) reconstructed InAs signatures at these sites, as it was observed in the STM studies here. However, despite these local dislocations the global trench symmetry always remains an (n×3) alignment, which is most likely due to the stable $In_{2/3}Ga_{1/3}$ stoichiometry within the InGaAs ML.

With the further deposition of InAs, the surface will become increasingly covered by the InAs-$\alpha 2/\beta 2(2\times 4)$ surface unit cells. However in this case, the trench dimers of the (4×3) reconstruction will not remain mostly unchanged as observed here due to their incommensurate threefold periodicity. In fact two thirds of the trenches must change their lattice position, and the corresponding As trench dimers from the (2×4) unit cells then must also form on top between Ga and In atom rows. If the underlying threefold periodicity of the group-III layer remains completely undisturbed during this process, the lattice periodicity of the fully covered second atomic layer consequently would correspond to a (2×12) surface unit cell.

From the presented structural model of the (2×4) reconstructed InAs islands the total amount of additional In accumulated on top of the $In_{2/3}Ga_{1/3}As$ ML can be estimated. Considering the number of six In atoms contained within each (2×4) unit cell and a total surface coverage of about 15 % this additional In amount is about 0.11 ML. Deducted from a total amount of 0.73±0.07 ML deposited on the surface of sample $\boxed{D}$, this leads to a nominal content of about 0.62±0.07 ML of pure InAs within the underlying (4×3) reconstructed layer, which nicely supports the assumption of a $In_{2/3}Ga_{1/3}$ stoichiometry in this layer. In turn, referring to the theoretically defined 0.67 ML InAs content in the underlying $In_{2/3}Ga_{1/3}As$ ML yields an actual total InAs content of 0.78±0.07 ML within both InAs layers on sample $\boxed{D}$, as shown in Tab. 7.2.

Following this discussion, the maximum InAs content of a fully evolved two layer InAs/InGaAs WL can be derived. Considering the content of 0.75 ML of InAs in a completed $\alpha 2/\beta 2(2\times 4)$ reconstructed ML and the 0.67 ML InAs content of the underlying $In_{2/3}Ga_{1/3}As$-(4×3) reconstructed ML the total InAs content of this WL amounts to 1.42 ML.

7.4.5. Summary

At an InAs coverage of about 0.67 ML the initial GaAs(0 0 1)-c(4×4) surface is dissolved and covered by an $In_{2/3}Ga_{1/3}As$-(4×3) reconstructed ML, for which a detailed structural model could be derived. The (4×3) surface unit cell is characterized by blocks of three As top dimers and a stable trench. Occasionally inserted η(6×3) and η(2×3) surface unit cells shift the alignment of the triple dimer blocks between in-line and brick-lined. Further deposited InAs is found to arrange in a (2×4) reconstruction on top of that $In_{2/3}Ga_{1/3}As$ ML in three slightly different configurations. Structural models for these $\alpha2$-(2×4), $\alpha2$-(2×4)-m and $\beta2$-(2×4) reconstructions are presented. These models are all in very well agreement with the observed STM data.

7.5. InAs thin films during quantum dot growth

7.5.1. Sample growth parameters

Sample E was covered with a material amount of 1.65 ML of InAs, typical for the critical thickness in QD growth [35, 68, 101]. The corresponding growth parameters are given in Table 7.4. During growth the transformation of the initial GaAs-c(4×4) reconstruction into the $In_{2/3}Ga_{1/3}As$ surface at about 0.67 ML of deposited InAs was observed by a clear change in the RHEED patterns. However, at about the critical thickness for QD growth (≈ 1.6 ML) the formation of diffraction spots in the RHEED patterns indicating the 3D transition could not be observed as clearly for this sample.

7.5.2. STM results

Overview

The STM image in Fig. 7.19 a provides a large scale view over the surface of sample E after the deposition of 1.65 ML of InAs onto the initial GaAs(0 0 1)-c(4×4) surface. As the deposited amount slightly exceeds the critical thickness for the 2D→3D transition, QDs have evolved. They appear as signatures of bright contrast and are found preferentially at the step edges of the surface terraces. For the QD signatures in Fig. 7.19 a the QD density is estimated to about $7.2 \cdot 10^{10}$ cm^{-2}.

In Fig. 7.19 b a surface area is displayed in more detail. Due to the high step density more QDs have assembled here, leading to a higher local QD density of about $9.3 \cdot 10^{10}$ cm^{-2} in this image.

The WL appears very clean, no significant contrast changes are observed on the terraces. It seems that all excess material is assembled in the QDs. Apart from the typical QDs

7.5. InAs thin films during quantum dot growth

		sample E
deposition time τ		170 s
calculated InAs deposition derived from the growth function		1.65 ± 0.07 ML
1st layer	coverage by (4×3) unit cells *	100% (by def.)
	estimated InAs content	0.67 ML (by def.)
2nd layer	coverage by (2×4) unit cells **	$85\pm5\%$
	estimated InAs content	0.64 ± 0.04 ML
QDs and precursors	QD density	$\approx 8.3 \cdot 10^{10}$ cm^{-2}
	precursor density	$\approx 1.6 \cdot 10^{10}$ cm^{-2}
	estimated InAs content	0.29 ± 0.05 ML
estimated total InAs content derived from the STM data		1.60 ± 0.09 ML

Table 7.4: *InAs growth parameters for sample* E *covered with an InAs amount slightly above the critical thickness for QD growth. The presented data relies on the STM findings in this work and follows the discussion in Appendix B.*
* *The surface area of one (4×3) surface unit cell corresponds to $A = 1.92$ nm^2.*
** *The surface area of one (2×4) surface unit cell corresponds to $A = 1.28$ nm^2.*

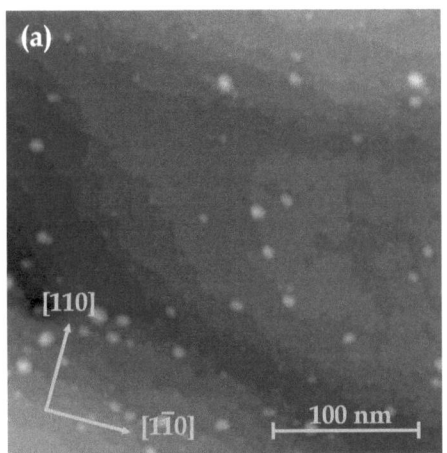
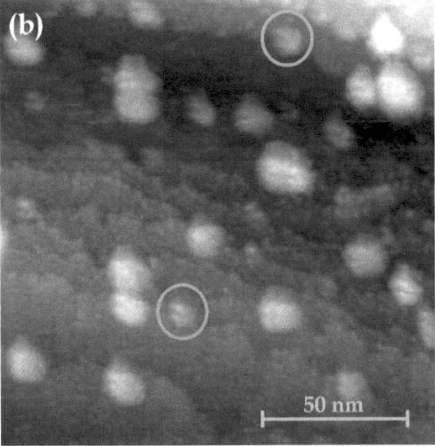

Figure 7.19: *Sample* E: *Filled state STM images ($V_T = -2.7$ V, $I_T = 0.2$ nA) of 1.65 ML InAs deposited on the GaAs(0 0 1) surface. (a) Large scale image providing an overview over the resulting surface. The surface appears smooth with larger terraces of monoatomic height. The bright signatures on top of the surface correspond to InAs QDs. They are preferentially located at surface step edges. (b) STM image with higher resolution focusing on the QDs. Due to the high step density, this particular area is characterized by a larger concentration of QDs. A tip-shape effect leads to a generally blurred contrast and an extended appearance of surface structures, which can most significantly be observed at the contrast of the QDs. Noticeable smaller InAs aggregations are marked by blue ovals.*

smaller InAs premature islands are found (marked by blue ovals), similarly located at the step edges, but with a smaller concentration of about four times less than the QDs.

Structural details of the QDs

The majority of the QDs observed here show a clear self-similarity in respect to their typical size. An average size of 13.5±1.4 nm in [$\bar{1}$ 1 0] direction, 10.3±1.0 nm in [1 1 0] direction, and 2.3±0.2 nm in height is yielded for a typical QD. This is in nice agreement with previous STM results on InAs/GaAs QD growth under comparable growth conditions [39].

The smaller premature islands exhibit about half the base size of the QDs and only show a height of about 2–3 ML. Thus it is assumed that these are QD precursors, e.g. small agglomerations of InAs that attract more material during growth at its site to fully evolve into QDs. The fast quenching of the sample directly after the growth was interrupted has probably then frozen these QD precursors in their current state.

However, QDs significantly larger than the typical size distribution were occasionally observed as well. In Fig. 7.20 a,b an STM image of such a QD is shown in detail. This QD is located directly above a surface step edge, which might induce its unusual, asymmetric, almost triangular shape. Investigations on QDs grown on tilted surfaces describe similar observations [172]. However, the influence of tip-shape effects that might also lead to such an STM contrast cannot be completely excluded.

From a height profile ① a base length of about 33 nm in [$\bar{1}$ 1 0] direction and a height of about 2.5 nm is derived (Fig. 7.20 c). The width is about 28 nm in [1 1 0] direction. The alignment of the appropriate surface directions is derived from the structural details of the WL. It should be noted that structural dislocations or relaxations within the structure of this relatively large QD are possible, but could not be detected in the present STM data.

The comparison of 3D views from this QD in Fig. 7.20 d with the well known atomically resolved InAs QD from Ref. [39] in Fig. 7.20 e shows some similarity of both QD structures. However, it is evident by the lack of contrast on the steep flanks of the QD investigated here that a large tip apex unfortunately corrupts the observed structural data. Such a tip attracts additional tunneling contributions especially at steep side flanks of the surface leading to superposition effects in the STM contrast, which become significant if the size of the apex is in the same range as the size of the observed structure. Thus the facets are not displayed in the same detail as in the reference image and a final statement on the possible facets of the present QD cannot be made.

Structural details of the WL

In the STM images in Fig. 7.21 the structure of the surrounding WL is shown in more detail. Two different types of structures are evident, corresponding to the two atomic layers discussed before. On the upper layer a (2×4) periodicity is observed (marked by blue arrows), with the unit cells often aligned in zig-zag chains. Thus the upper layer is ascribed

7.5. InAs thin films during quantum dot growth

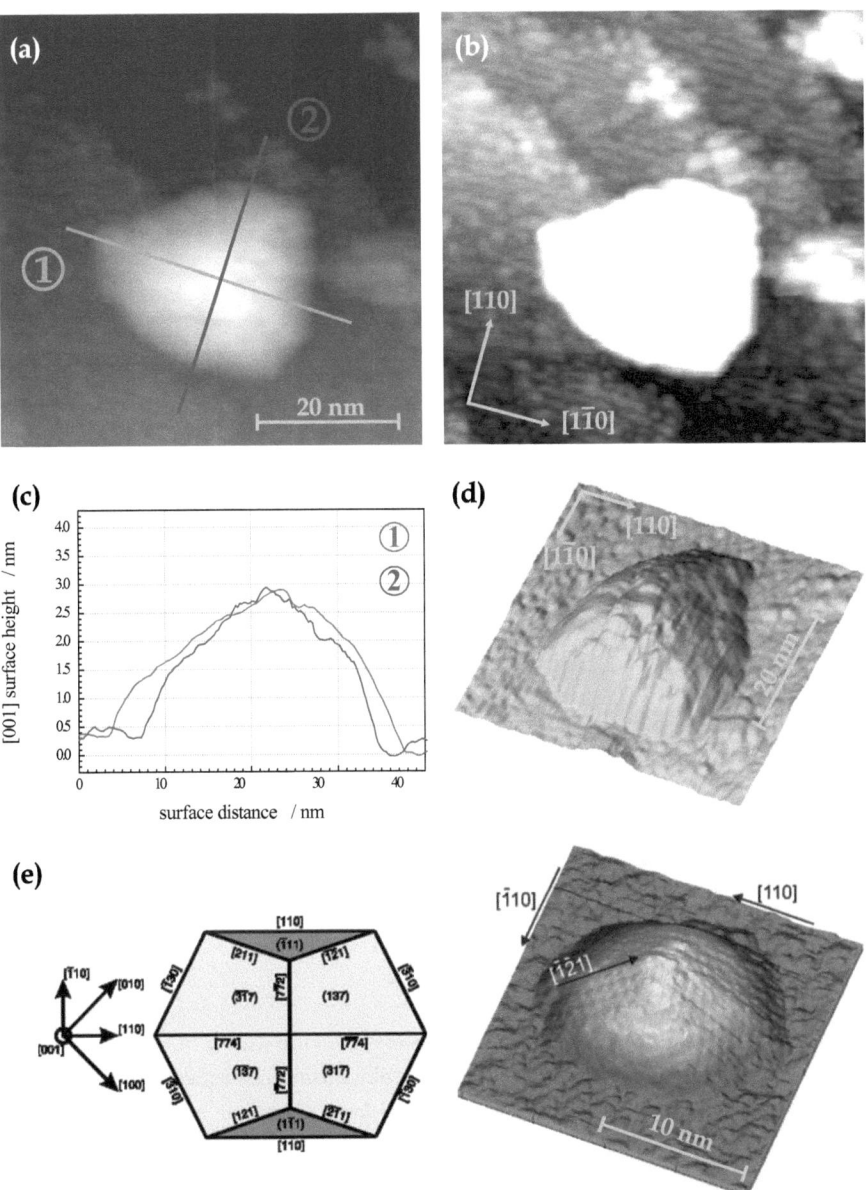

Figure 7.20: Sample E: *(a) Filled state STM image ($V_T = -2.7\,V$, $I_T = 0.2\,nA$) of one of the comparatively larger QDs. (b) The contrast is adjusted to the WL structure in order to assign the appropriate surface directions. The data of the height profiles ① and ② are shown in (c). In (d) a 3D view of the STM data is shown in comparison to a reference image (e) of the atomically resolved InAs QD shape as presented in Ref. [39] and the established structural model of the InAs QD as presented in Ref. [40].*

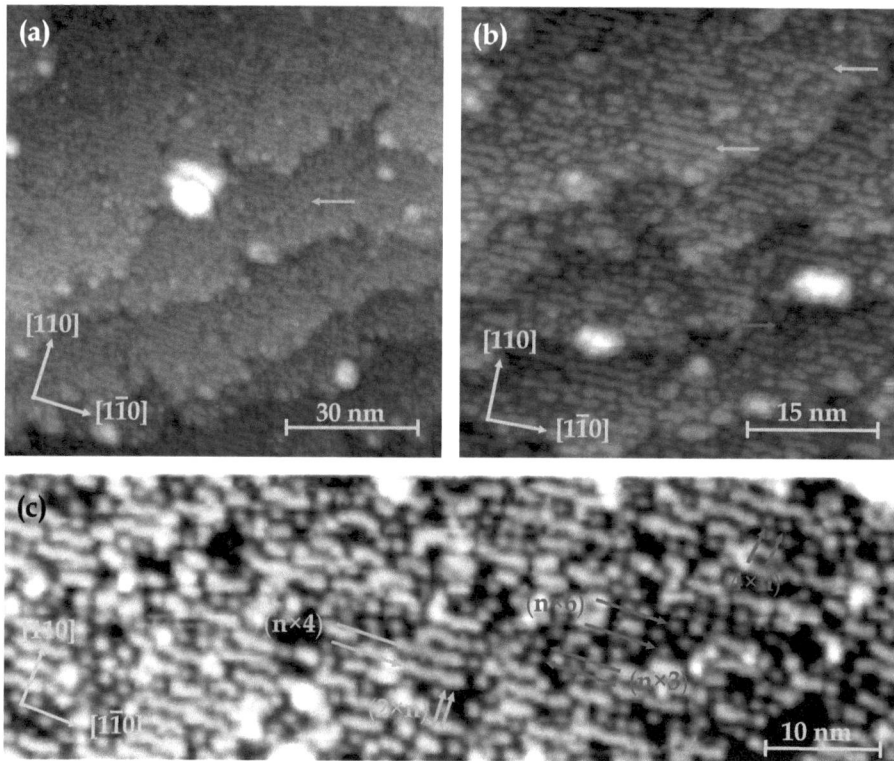

Figure 7.21: *(sample E) Atomically resolved filled state STM images ($V_T = -2.7\,V$, $I_T = 0.2\,nA$) of the WL. The WL apparently consists of two layers, on which two different reconstruction patterns are observed. The upper layer (marked by blue arrows) is congruent to the InAs-(2×4) reconstructions as already discussed. This layer does not cover the entire surface. The locally uncovered layer underneath (marked by magenta arrows) is congruent to the InGaAs-(4×3) reconstruction. In (c) the corresponding surface periodicities are exemplarily shown in more detail.*

to the InAs-(2×4) reconstructions that were observed on the second InAs ML before. However, the atomic resolution does not suffice to locally distinguish between the $\alpha 2/\alpha 2$-m and $\beta 2$ configuration. The upper (2×4) reconstructed layer covers the majority of the surface, roughly estimated to about 85%. About 15% of the underlying surface is left uncovered. Here mostly a (4×3) reconstruction is observed (marked by magenta arrows), but occasionally the (4×6) reconstruction is found as well. Consequently this layer is ascribed to the $In_{2/3}Ga_{1/3}As$-(4×3) reconstructed ML.

The alignment of the surface unit cells appears slightly deformed due to surface drift effects while the STM image was taken. However, the limited atomic resolution does not allow to appropriately correct the image.

7.5.3. Discussion: The InGaAs/InAs wetting layer during quantum dot formation

Following the results on the InAs ML discussed before, a further InAs deposition on the underlying $In_{2/3}Ga_{1/3}As$-(3×3) surface layer leads to the overgrowth by an (2×4) reconstructed InAs layer. This is in nice agreement with the STM results found here.

The deposited material amount evidently satisfies the critical thickness for the 2D→3D transition, as QDs can be clearly observed. The size and shape of most of these QDs are in good agreement with previous studies on InAs/GaAs QD growth under comparable growth conditions [39,191]. For a deposited amount of about 1.8 ML InAs the QD density in Ref. [39] is reported to be about $1.9 \cdot 10^{11}$ cm^{-2}, which is significantly larger that the QD density of 7.2–$9.3 \cdot 10^{10}$ cm^{-2} derived here, even considering a correction factor of about 1.2 to account for the five times less observed smaller QD precursors. Consequently the actually deposited amount of InAs on sample E here is less than in Ref. [39], being consistent with the nominal value of 1.65 ML derived from the growth function in Appendix B.

From the average QD size and density the amount of InAs assembled in the QDs and precursors is estimated to about 0.29 ± 0.05 ML in total. Furthermore it is observed that the (2×4) reconstructed coverage at the InAs WL is not fully closed, covering roughly about 85% of the WL surface. Considering this, the total amount of InAs assembled in the WL is about 1.31 ± 0.04 ML. In result, a total amount of InAs of 1.60 ± 0.09 ML deposited during growth is derived, in nice agreement with the nominally deposited amount.

In comparison to previous reports on InAs/GaAs QD growth, a deposited amount of about 1.5 ML pure InAs is considered the lowest threshold for the critical thickness [35]. However, a significant population of QDs, which showed evidence in the RHEED diffraction patterns, was only observed for slightly larger InAs quantities of about 1.6 ML [35,68] using comparably slow growth rates between $0.01-0.02$ ML/s and of about 1.75 ML [101] using a comparatively faster growth rate of 0.06 ML/s.

In the results of this work, at the very slow growth rate of 0.007 ML/s, QD related RHEED pattern were not clearly observed during growth. This is presumably due to the low amount of deposited InAs of about 1.65 ML, very close to the 3D transition threshold, and the rapid quenching immediately after deposition.

The 3D transition certainly is not an instant event, but more an extended phase within an evolutionary process. As the InAs density on the surface increases, QD precursor sites are formed, which then evolve to QDs. During the growth here, the evolved number of QDs at 1.65 ML deposited InAs probably was not sufficient to significantly contribute to the RHEED diffraction patterns. Moreover, it is known that the post-growth sample treatment has a large influence on the further evolution of the QDs as well. An annealing phase after growth induces a ripening of the QDs at the expense of smaller QDs and material from the WL. The rapid quenching of samples has been proven to minimize such effects [102].

However, at low deposition rates a QD ripening may also yet occur during growth. As discussed before, the WL in the investigations here exhibited an incomplete coverage by the second InAs layer with a deficit of about 15% or 0.11 ML. Estimating the InAs amount assembling in the QDs from Ref. [39] in a similar way, a value of 0.63 ± 0.05 ML is yielded. Considering the total amount of 1.8 ML of deposited InAs in Ref. [39], this would leave a deficit of 0.25 ML or 33% within the second InAs layer of the WL, indicating a QD ripening consuming InAs from the WL. The STM data in Ref. [39] unfortunately does not allow to support or decline this consideration.

Yet, it may be presumed from these findings and the following discussion in Sect. 7.6 that the complete second layer coverage during the WL evolution (1.42 ML InAs) accumulates unfavorable amounts of strain energy. The deposited InAs thus starts to assemble in precursor states that evolve to QDs during further growth. During further growth the additional InAs accumulating in the QDs does not only originate from the further impinging material but must also originate from material already deposited in the WL. The critical thickness for the 2D→3D transition then corresponds to a fully evolved two-layer WL containing 1.42 ML InAs, where the accumulated strain energy is the driving force for a material relocation from the WL to the QDs. However, as some material may already be assembled in precursor states, the observed critical thickness could eventually be slightly larger.

7.5.4. Summary

At a total amount of 1.65 ML InAs deposited on the initial GaAs(0 0 1) surface, typical QDs were observed with a density of about $7.2 - 9.3 \cdot 10^{10}$ cm^{-2}. In contrast, smaller QD precursors were observed with an additional share of only 20%. The WL was observed to be partly incomplete, missing about 15% of the second InAs-(2×4) reconstructed layer. This corresponds to a missing amount of 0.11 ML InAs that probably accumulated in the QDs. The total InAs amount in the QDs and precursors is estimated to about 0.29 ML.

From these results the following growth model is derived. The increasing deposition of InAs onto the In$_{2/3}$Ga$_{1/3}$As-(4×3) reconstructed layer leads to a further coverage by a second (2×4) reconstructed InAs layer. At a completed two layer WL, containing 1.42 ML, the critical thickness for the 2D→3D transition is approached, and material from the WL becomes relocated into QD precursor states due to strain relief providing attractive sites for further InAs accumulation. During this process, the precursors evolve to QDs (*ripening*) by accumulating material from the WL as well as from the molecular beam. A significant density of QDs must evolve before they can visibly contribute to the diffraction patterns in e.g. RHEED. The fast quenching of the sample immediately after material deposition interrupts this process and both QDs as well as precursors can be observed, even though there might not be clear evidence for QDs in the diffraction patterns.

7.6. Strain effects during the evolution of the InAs wetting layer

In the following, the different evolutionary stages during the InAs thin film growth on the GaAs(0 0 1)-c(4×4) surface will be discussed in detail, in particular with respect to the bond lengths of the surface atoms in the respective structural models [192].

7.6.1. Strain at the GaAs(0 0 1)-c(4×4) surface

Considering the bare GaAs-c(4×4) reconstruction each of the As top dimers, which form the characteristic triple dimer blocks, is attributed with an As–As bond length of only about 0.26 nm [77], considerably less than the expected distance of 0.40 nm for the two As atoms at their respective lattice positions before dimer formation (Fig. 7.22). This results in a tensile strain of the As dimer along the dimer bond direction, corresponding to the [1 1 0] direction. Additionally the triple As dimers are chemisorbed on top of a pure As layer with a

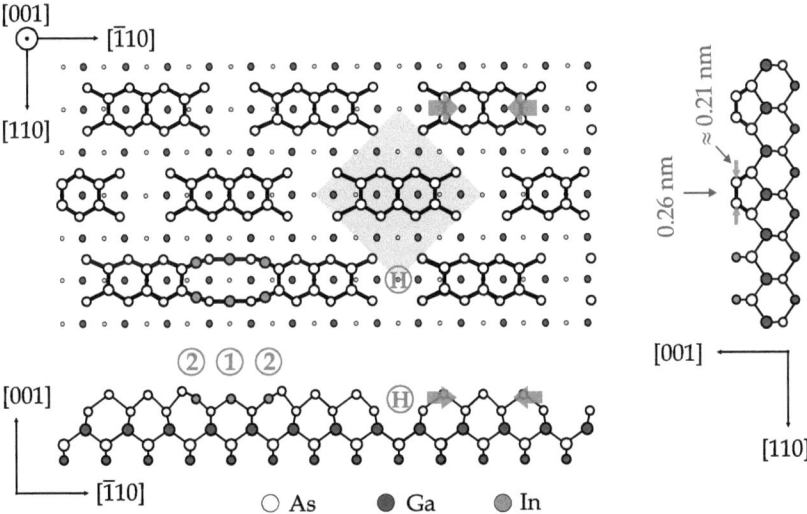

Figure 7.22: Structural model depicting the strain situation at the undisturbed GaAs(0 0 1)-c(4×4) surface in top and side views. The model illustrates the structure of the initial empty hollow sites and a hollow site occupied by one of the InAs signatures. The averaged values of the As bond lengths denoted in the side view are derived from Ref. [77]. The respective strain effects are illustrated by green arrows. The surface unit cell is marked by a yellow box. Atoms below the figure plane are depicted by smaller circles.

backbond length of about 0.21 nm between the dimer atoms and the atoms underneath [77]. This is considerably shorter as compared to the bulk related Ga–As bond length of 0.243 nm and thus leads to additional tensile strain in the [$\bar{1}$ 1 0] direction and an in plane shift of the underlying As atoms towards the center dimer position so that the As triple dimers are shifted towards each other. The strain relaxation in the [$\bar{1}$ 1 0] direction results in the formation of a dimer vacancy at every fourth As dimer position (hollow sites). This is illustrated by green arrows in Fig. 7.22. In the [1 1 0] direction, the trench between the As triple dimer rows prevents the surface from accumulating significant surface strain.

7.6.2. Strain related to the InAs signatures on the GaAs-c(4×4) surface

In contrast, the incorporation of In atoms into their designated bulk positions in the center of the hollow site, marked by ① in Fig. 7.22, causes a moderate compressive strain along the [$\bar{1}$ 1 0] direction because of the larger In–As bond length (in bulk 0.262 nm) as compared to the Ga–As bond length (in bulk 0.243 nm). This partly compensates the tensile strain at the As triple dimer blocks. However, over a distance of four surface lattice unit cells along the [$\bar{1}$ 1 0] direction two In–As bonds with a length factor of 1.08, as compared to the GaAs bulk, will have to compensate six As–As backbonds with a bond length factor of only 0.85:

$$(2 \cdot 1.08 + 6 \cdot 0.85) / (2 + 6) = 0.91 \qquad (7.4)$$

This tensile strain can be reduced by the incorporation of further four In atoms within the outer dimer As–As backbonds, marked by ② in Fig. 7.22. This increases the contribution of the compressive strain and may thus effectively compensate the tensile strain at the As triple dimer blocks. Additionally, as the trench between the rows of triple dimer blocks remains stable during the In incorporation, so that the In-induced strain accumulation along the [1 1 0] direction remains insignificant as well.

7.6.3. Strain at the $In_{2/3}Ga_{1/3}As$-(4×3) reconstructed monolayer

The $In_{2/3}Ga_{1/3}As$-(4×3) reconstructed surface, as described in the structural model as shown in Fig. 7.23, is built from three atomic layers. First there is a group-III terminated atomic layer with a threefold periodicity of one row of Ga atoms and two rows of In atoms alternating in the [1 1 0] direction which causes an inhomogeneous compressive strain along this direction. The slightly larger atomic radii and backbond lengths of the In atoms — as compared to the Ga atoms — further result in a slight buckling of this layer, affecting the second — As terminated — atomic layer as well (illustrated by green arrows in the right side view of Fig. 7.23). In the second layer the As trench dimers above the In atom rows are relaxed outwards of the surface as compared to their designated bulk positions, while the As atoms above the Ga atom rows are slightly shifted inwards to the center of the Ga atoms in [1 1 0] direction. The As dimer blocks of the third layer then form above the Ga related As atom rows as this shift allows the dimer bonds to relax towards their natural bond lengths. The compressive strain in [1 1 0] direction caused by the group-III atoms then is compensated by the tensile strain of the As top dimers and the tensile strain caused by the As trench dimers, that pulls the backbonded In atoms to the dimer bond axis.

Along the [$\bar{1}$ 1 0] direction compressive strain is accumulated in the In atom rows on the GaAs bulk lattice. On the other hand, the shorter As–As backbonds of the As dimer blocks cause tensile strain, by which the As dimers are shifted together and dimer vacancies (*hollow sites*) form at about every fourth dimer position to relax this strain. Both processes are illustrated by arrows in the lower side view of Fig. 7.23. Over a distance of four surface lattice unit cells along the [$\bar{1}$ 1 0] direction then eight In–As bonds with a length factor of

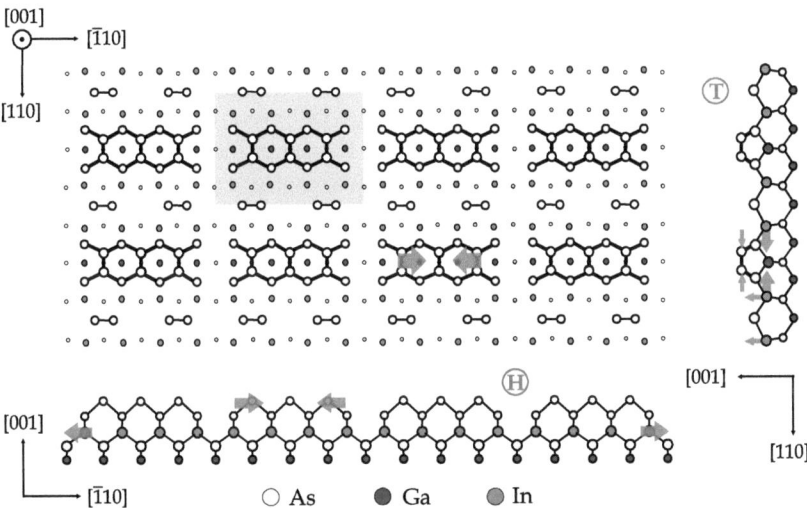

Figure 7.23: Structural model of the $In_{2/3}Ga_{1/3}As$-(0 0 1)-(4×3) surface depicting the strain situation at the undisturbed surface in top and side views. The strain induced hollow site between the As dimer blocks is marked by Ⓗ, the characteristic As trench dimer site is marked by Ⓣ. The yellow box illustrates the undisturbed surface unit cell. In [$\bar{1}$ 1 0] direction the tensile strain at the triple As dimer blocks caused by the shorter bond length is almost completely compensated by the compressive strain caused by the lattice mismatched chemisorption of the slightly larger In atoms onto the GaAs bulk. In [1 1 0] direction — due to the $In_{2/3}Ga_{1/3}$ stoichiometry — every third group-III atom is a Ga atom allowing the neighboring In atoms to partly relax towards the Ga bulk positions. The slightly shorter distance between adjacent As atoms on top of the Ga atoms is favorable for the formation of As–As top dimers as it fits well to their shorter bond length. At the trench dimer site, the In atoms may also slightly relax outwards the surface. The respective strain effects are illustrated by green arrows.

1.08 — as compared to the GaAs bulk — effectively compensate six As–As backbonds with a bond length factor of 0.85:

$$(8 \cdot 1.08 + 6 \cdot 0.85) / (8 + 6) = 0.98 \tag{7.5}$$

This effective strain compensation solely due to the different bond lengths on the surface may also be responsible for the imperfect periodicity of the top-dimer/hollow-site alignment along the [$\bar{1}$10] direction as compared to the GaAs-c(4×4) surface. There the hollow sites provide the only possibility for a strain reduction of the As triple dimers and are thus considerably large and stable. For the hollow sites on the $In_{2/3}Ga_{1/3}As$-(4×3) surface the strain to compensate is much less (cf. Eqs. 7.4 and 7.5) and thus they were observed to be not as stable. Finally, it is assumed that the trench dimer only very moderately contributes to the tensile strain in [$\bar{1}$10] direction. DFT calculations revealed this for the trench dimers on the InAs-α2(2×4) reconstruction [193], which are quite similar to the trench dimers here.

7.6.4. Strain at the InAs-(2×4) reconstructed second layer

The further coverage of the $In_{2/3}Ga_{1/3}As$ ML leads to the formation of (2×4) reconstructed InAs signatures on top. To form this additional InAs layer the topmost As dimers of the underlying (4×3) reconstructed $In_{2/3}Ga_{1/3}As$ must be removed and thus will not contribute to the strain any more. On the other hand, tensile strain is accumulated along both surface

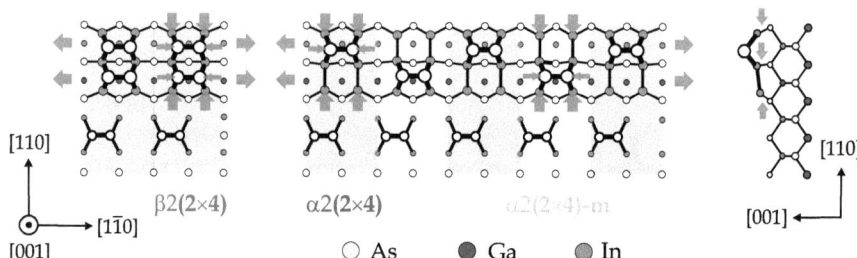

Figure 7.24: Structural model of the InAs-(2×4) reconstructed layer on top of the $In_{2/3}Ga_{1/3}As$-(001)-(4×3) surface depicting the surface strain situation in top and side views. At the β2(2×4) configuration, the comparatively short As–In backbond length of the As dimers causes a moderate tensile strain, partly compensating the compressive strain in [110] direction originating from the $In_{2/3}Ga_{1/3}$ layer underneath. The formation of In–In dimer bonds at the α2/α2-m configuration even intensifies the tensile strain in [110] direction and thus the compensation of the compressive strain resulting from the underneath layer. In [$\bar{1}$10] direction the tensile strain from the As top dimers cannot sufficiently compensate the compressive strain caused by the comparatively larger In atoms. Thus with the increasing (2×4) coverage, a significant amount of strain energy is accumulated, finally resulting in QD growth. The respective strain effects are illustrated by green arrows.

main directions by the evolving InAs-(2×4) reconstructions, illustrated by green arrows in Fig. 7.24.

In the [1 1 0] direction this tensile strain is caused by the As top dimers and trench dimers due to their short As–In backbond lengths. Furthermore — considering the $\alpha 2(2\times 4)$ configuration — additional In–In bonds characterized by a rather short bond length are formed. As calculated for the pure $\alpha 2(2\times 4)$ surface this In–In bond length of 0.278 nm has a bond length factor of only 0.70 as compared to the distance of two As atoms of the underlying GaAs surface in the same direction [193]. Both effects thus allow compensation of the compressive strain contributions from the $In_{2/3}Ga_{1/3}$ layer underneath. However, in the [$\bar{1}$ 1 0] direction the As dimers can only provide a moderate contribution to the tensile strain.

It may be assumed that the situations of the total strain of the As top dimers and the In–In bonds are different. In this case an alternating occurrence of the $\alpha 2(2\times 4)$ and the $\alpha 2(2\times 4)$-m unit cells would be favorable, resulting in a zig-zag chain alignment of unit cells along the [$\bar{1}$ 1 0] direction (Fig. 7.24). This in fact was mostly observed in the STM images. The $\beta 2(2\times 4)$ unit cells, on the other hand, would not apply to such an arrangement. Even though the formation of additional As dimers instead of the In–In bonds reduces some strain, the $\beta 2(2\times 4)$ is less often observed.

Eventually, the (2×4) reconstructed InAs signatures on the $In_{2/3}Ga_{1/3}As$ surface exhibit a much larger compressive strain along the [$\bar{1}$ 1 0] direction as compared with the c(4×4) and (4×3) reconstructed structures discussed before.

7.6.5. Strain during quantum dot formation

The previously discussed surface configurations of the three stages during the InAs WL evolution on GaAs(0 0 1) all show their ability to compensate accumulated strain energy. Yet, already in the third stage a significant amount of compressive strain remains along the [$\bar{1}$ 1 0] direction at a complete coverage by the InAs-(2×4) reconstructed unit cells. At this point the WL contains 1.42 ML InAs. Any further InAs layer growth would rapidly induce more compressive strain that cannot be compensated sufficiently. In this situation QD growth is favorable. The transition from 2D to 3D growth allows to sufficiently relax the accumulated compressive strain in particular close to the apex of the QDs. The broken surface symmetry at surface steps may also allow an additional lateral strain relaxation, which could explain the observed preferential occurrence of QDs at these sites. Furthermore, the strain relaxation in the QDs seems favorable even for material from the WL, as the experimental findings suggest a material relocation from the partly strained (2×4) reconstructed upper layer of the WL into the QDs.

7.6.6. Summary

The proposed structural models for the three stages of the WL evolution in InAs/GaAs heteroepitaxy were discussed with regard to the surface strain situation, indicating a considerable compensation of strain energy at the different growth stages.

In the first stage of the WL evolution InAs is adsorbed at the hollow sites of the initial GaAs-c(4×4) surface, as vacant lattice positions are offered here. Though the atomic structure of the adsorbed InAs is not yet known in full detail, the proposed structural model for its incorporation into the hollow sites effectuates a significant strain relaxation and leads to a low strain situation along both surface directions.

With the transformation into the $In_{2/3}Ga_{1/3}As$-(4×3) reconstructed surface, the strain situation becomes more relaxed. The formation of differently aligned As dimers effectively compensates the buckling caused by the mixed group-III layer and the compressive strain from the incorporation of the slightly larger In atoms. Thus the $In_{2/3}Ga_{1/3}As$-(4×3) reconstruction of the second stage of the WL evolution is characterized by a low strain situation along both surface directions as well.

With the formation of (2×4) reconstructed unit cells in the third stage of the WL evolution, strain can be sufficiently reduced only along the [1 1 0] direction. Along the [$\bar{1}$ 1 0] direction, in contrast, considerable compressive strain is accumulated with an increasing coverage by the (2×4) reconstructed unit cells.

At the completed third stage, corresponding to a fully evolved two-layer WL with an InAs content of 1.42 ML, any further adsorbed InAs layer would rapidly increase the compressive strain unable to be sufficiently compensated. This strain situation leads to a 2D→3D transition during growth and thereby to the formation of QDs, where the compressively strained InAs material can relax more efficiently. Further deposited InAs then solely assembles in the QDs at this stage. Moreover, some material from the upper layer of the partly strained WL is relocated into the QDs as well.

8. InAs thin film growth on GaAs(0 0 1)-$\beta 2(2\times 4)$

8.1. Introduction

The growth regime on the GaAs(0 0 1)-$\beta 2(2\times 4)$ reconstructed substrate surface is commonly used especially for the GaAs homoepitaxy [194]. Furthermore it is an important regime for the growth of sophisticated nanostructures, e.g. stacked QDs and large QDs for the long wavelength range [12,82]. Typical growth temperatures between 510–530 °C ensure a high mobility of deposited species at the surface, allowing the ripening of larger QDs and a larger uniformity of these structures. Yet, for τ_c, the critical time of the 2D→3D transition during InAs deposition, a significant delay is reported as compared with the growth regime on the GaAs(0 0 1)-c(4×4) surface at substrate temperatures below 480 °C [29]. A larger τ_c may lead to the assumption of a larger actual critical thickness resulting as well. However, the delay for τ_c is mainly attributed to InGaAs alloying and the increased In desorption from the surface at higher substrate temperatures [29].

8.2. Experimental details

8.2.1. Sample preparation

In principle the preparation of the samples followed the same scheme as applied in the case of the previously discussed samples. Yet, after buffer layer growth and annealing, the sample temperature was maintained at a level, where the (2×4) RHEED patterns of the GaAs(0 0 1)-$\beta 2(2\times 4)$ reconstruction could be clearly observed. The corresponding sample growth scheme is illustrated in Fig. 8.1.

In preparation for the InAs deposition, the sample was annealed for another 10 min at a substrate temperature $T_S = 530$ °C to ensure a stable $\beta 2(2\times 4)$ reconstructed surface. Then InAs was deposited for a time t_{dep} at an As/In flux ratio of about 130. To avoid the transition of the GaAs substrate surface from the $\beta 2(2\times 4)$ to the more As rich c(4×4) reconstruction during quenching, the As_2 supply was interrupted immediately after deposition while T_S was kept at the 530 °C level for another 30 s. Subsequently the sample was quenched rapidly below 327 °C in less than 2 min to literally freeze its current state, while RHEED was used to control that the surface structure did not change during this process. The sample was then transferred *in situ* into the STM chamber for structural analysis.

8.2.2. Estimation of the growth rate

In contrast to the InAs deposition at lower substrate temperatures, the desorption of InAs from the surface back into the residual gas becomes significant during growth in the temperature regime of the GaAs(0 0 1)-β2(2×4) reconstruction. At $T_S = 530$ °C the desorption process of InAs is expected at a rate of more than 0.0065 ML/s [195], which is significant in comparison to the estimated InAs deposition rate of roughly about 0.010 ML/s ($BEP_{In} \approx 2.7 \cdot 10^{-8}$ mbar). The significance of the desorption process complicates a reliable prediction of the actual InAs growth rate. Considering both values the resulting InAs growth rate can be assumed in a first simple approximation to about 0.0035 ML/s.

However, since the shutter of the In K-cell was inoperative, a time dependent deposition rate results. Furthermore, during the 30 s growth interruption before quenching, an additional amount of about 0.2 ML InAs is expected to desorb from the surface as derived from the InAs desorption rate. A more detailed discussion and the resulting actual growth function are given in Appendix B.

8.3. InAs thin films on GaAs(0 0 1)-β2(2×4)

8.3.1. Sample growth parameters

In the following the experimental findings on three samples are presented, systematically prepared with an increasing amount of InAs according to the general growth scheme from Fig. 8.1. The corresponding growth parameters are given in Tab. 8.1.

During growth of each sample, RHEED was monitored, yet no evident changes in the diffraction patterns were observed during growth. However, the RHEED observations in these experiments suffered from a slight blurring of the diffraction pattern. Thus, only the initial β2(2×4) surface periodicity and the constancy of patterns during quenching can be guaranteed here. Beyond this, minor changes in the RHEED diffraction patterns during growth may not have been detected.

8.3.2. STM results

Overview

The STM images in Figs. 8.2, 8.3, and 8.4 provide an overview on the different samples after the respective InAs deposition given in Tab. 8.1. On a large scale the surface of each sample appears very smooth with large flat terraces separated by steps of monoatomic height.

On sample F an estimated amount of 0.31 ML of InAs is deposited onto the initial GaAs(0 0 1)-β2(2×4) reconstructed surface. The corresponding STM image in Fig. 8.2 shows signatures of bright contrast distributed well across the surface. Similar to the previous findings on the GaAs(0 0 1)-c(4×4) reconstructed surface, these signatures are assumed to be related to the deposited InAs. They appear very well aligned along the [$\bar{1}$ 1 0] direction,

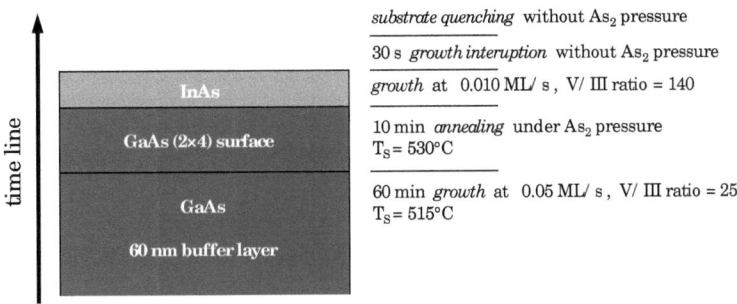

Figure 8.1: *General sample growth scheme for the preparation of InAs thin films on the GaAs(0 0 1)-β2(2×4) surface, while the amount of deposited InAs was varied systematically.*

		sample F	sample G	sample H
deposition time τ		170 s	200 s	260 s
calculated InAs deposition derived from the growth function		0.31±0.03 ML	0.51±0.03 ML	0.92±0.03 ML
1st layer	coverage of (2×4) unit cells *	32±5%	45±5%	-
	coverage by (4×3) unit cells **	-	-	100% *(by def.)*
	estimated InAs content	0.32±0.05 ML	0.45±0.05 ML	0.67 ML *(by def.)*
2nd layer	coverage by (2×4) unit cells *	1.6±0.5%	4.2±0.5%	31±5%
	estimated InAs content	0.01±0.01 ML	0.03±0.01 ML	0.23±0.04 ML
estimated total InAs content derived from the STM data		0.33±0.05 ML	0.48±0.06 ML	0.90±0.04 ML

Table 8.1: *InAs growth parameters for samples F–H grown on the initial GaAs(0 0 1)-β2(2×4) surface. The presented data relies on the STM findings and follows the discussion in Appendix B.*
* The surface area of one (2×4) surface unit cell corresponds to $A = 1.28\ nm^2$.
** The surface area of one (4×3) surface unit cell corresponds to $A = 1.92\ nm^2$.

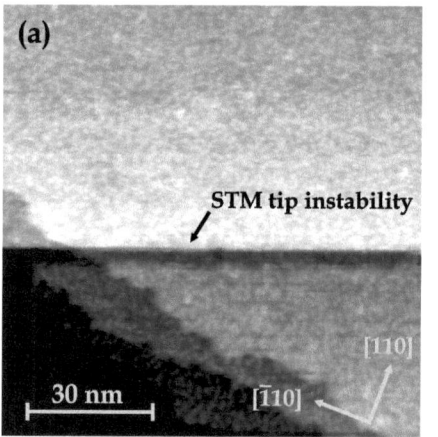 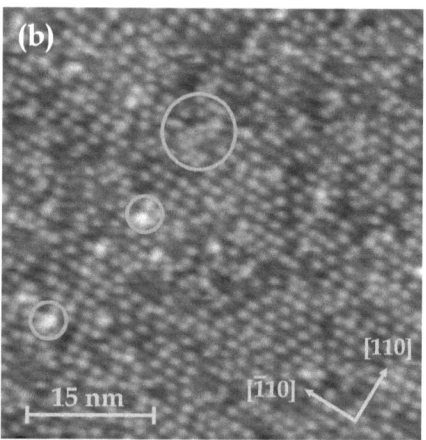

Figure 8.2: Sample ⟦F⟧: Filled state STM images ($V_T = -3.9\,V$, $I_T = 0.1\,nA$) of the GaAs(0 0 1)-$\beta 2(2\times 4)$ reconstructed surface after the deposition of about 0.31 ML of InAs. (a) The large scale image shows a smooth surface with flat terraces separated by steps of monoatomic height. During imaging, a significant change in the STM contrast occurred due to a tip instability. The resulting contrast change appears very similar to the contrast change caused by surface steps. (b) In a more detailed view signatures corresponding to InAs appear well distributed throughout the surface in a typical alignment clearly separated by an apparent trench in [1̄ 1 0] direction. However, as marked by blue circles, rarely a much more compact alignment of the signatures is observed. Some very bright contrast is also observed and ascribed to InAs adsorbed on top of the signatures, as marked by green circles.

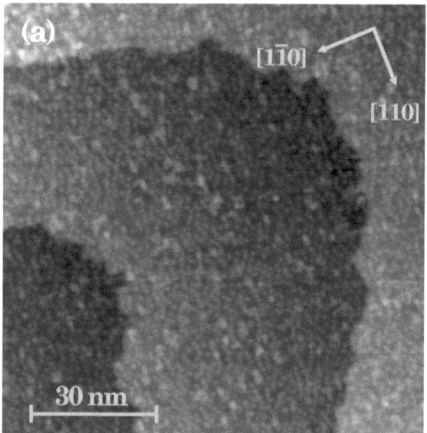 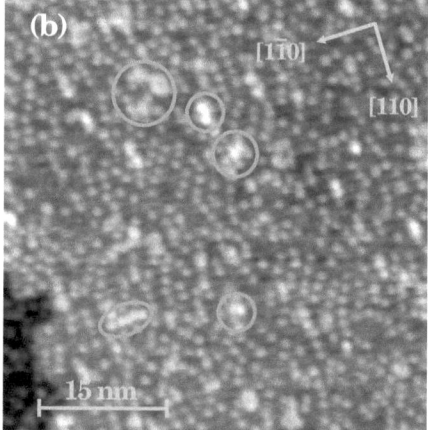

Figure 8.3: Sample ⟦G⟧: Filled state STM images ($V_T = -2.7\,V$, $I_T = 0.7\,nA$) of the GaAs(0 0 1)-$\beta 2(2\times 4)$ reconstructed surface after the deposition of about 0.51 ML of InAs. (a) The large scale image shows a smooth surface with flat terraces separated by steps of monoatomic height. (b) In a more detailed view the InAs signatures evenly cover the complete surface, separated by the trench in [1̄ 1 0] direction. The more compact alignment observed before on sample ⟦F⟧ could not be observed here. The bright contrast ascribed to InAs adsorbed on top of the signatures is marked by green circles.

8.3. InAs thin films on GaAs(0 0 1)-β2(2×4)

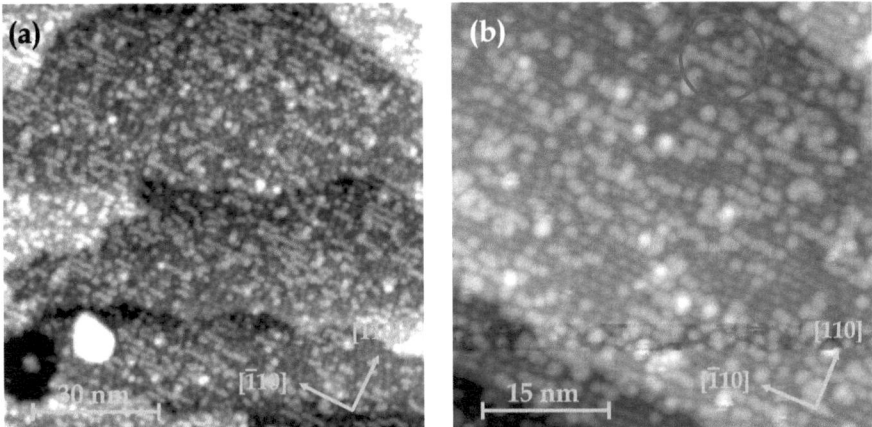

Figure 8.4: Sample [H]: Filled state STM images ($V_T = -3.3\,V$, $I_T = 0.2\,nA$ and $V_T = -4.3\,V$, $I_T = 0.3\,nA$) of the GaAs(0 0 1)-β2(2×4) reconstructed surface after the deposition of about 0.92 ML of InAs. (a) The large scale image shows a surface with flat terraces and an apparently well ordered alignment of InAs signatures. (b) The more detailed view reveals a surface configuration different from the signature alignment observed before on samples [F] and [G]. There is a well ordered surface reconstruction underneath and bright InAs related islands on top, that preferably align in (zig-zag) chains along the [$\bar{1}$ 1 0] direction (exemplarily marked by a magenta circle).

with a similar size and mostly characterized by an equal intermediate distance. However, occasionally the signatures are found in a much closer packaging, directly neighboring each other (exemplarily marked by a blue circle in Fig. 8.2 b). Additionally some much brighter contrast is observed (marked with green circles in Fig. 8.2 b) probably corresponding to additional InAs on top of the other signatures, partly forming a second layer.

Similar InAs signatures were also found on sample [G] after the deposition of about 0.51 ML of InAs. The signatures exhibit a similar alignment with similar intermediate distances, as shown in Fig. 8.3. However, the much closer packaging of signatures could not clearly be observed any more. Yet, the much brighter contrast from the second layer InAs (marked exemplarily by green circles in Fig. 8.2 b) was observed much more often.

In contrast to the two samples discussed before, there is a clear change of the surface configuration observed on sample [H] after the deposition of about 0.92 ML of InAs (Fig. 8.4). There are signatures of bright contrast strictly aligned in rows, which form zig-zag chains very similar to the InAs-(2×4) structures observed on sample [D] (exemplarily marked with a magenta circle in Fig. 8.4 b). The uniform distribution of single signatures as observed on samples [F] and [G] is not observed here any more. Furthermore, the underlying surface exhibits a reconstruction that can be clearly distinguished from the initial GaAs(0 0 1)-β2(2×4) reconstructed surface (cf. Fig. 6.11 a).

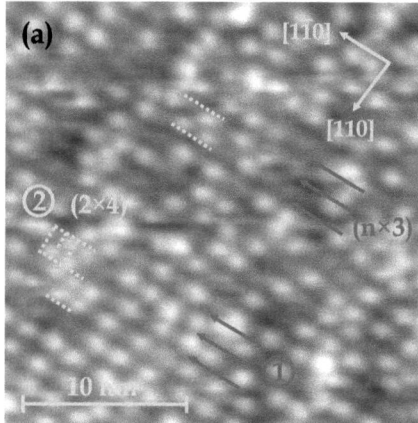 

Figure 8.5: (a) Filled state STM image from sample $\boxed{F}$ ($V_T = -3.9\,V$, $I_T = 0.1\,nA$) showing the initial GaAs(0 0 1)-β2(2×4) reconstructed surface after the deposition of about 0.31 ML of InAs. The typical alignment of the signatures exhibits an (n×3) periodicity which is incommensurate to the initial (2×4) reconstruction ①. Only rarely the more compact alignment of signatures is observed, corresponding to (2×4) surface unit cells (marked by dotted yellow boxes ②). Each of the signatures can be assigned to one (2×4) surface unit cell, yet in alternating zig-zag alignment of adjacent signatures. (b) Filled state STM image from sample $\boxed{G}$ ($V_T = -2.7\,V$, $I_T = 0.7\,nA$) after the deposition of 0.51 ML of InAs. The threefold periodicity of the signature alignment in [1 1 0] direction is again evident ③. A fourfold periodicity in [$\bar{1}$ 1 0] direction ④, leads to a typical alignment corresponding to (4×3) or c(4×6), (illustrated by yellow boxes). The number of bright contrast features ascribed to InAs adsorbed on top of the signatures has clearly increased.

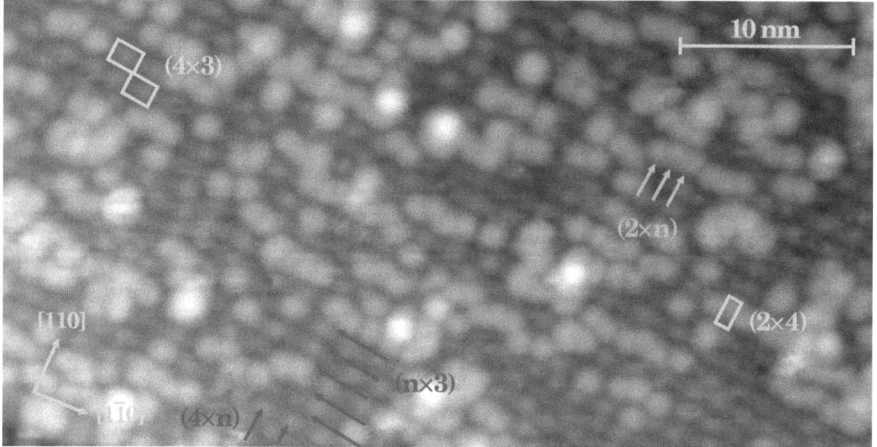

Figure 8.6: Filled state STM image from sample $\boxed{H}$ ($V_T = -4.3\,V$, $I_T = 0.3\,nA$) after the deposition of about 0.92 ML InAs onto the GaAs(0 0 1)-β2(2×4) surface. The surface reconstruction has transformed into a clear (4×3) reconstruction and additional InAs is adsorbed in small (2×4) islands on top of that layer. The corresponding surface unit cells are marked by yellow boxes. The observed surface periodicities are exemplarily marked by magenta and blue arrows.

Structural details

In the STM image in Fig. 8.5 a the surface of sample ⟦F⟧ covered by an InAs amount of about 0.31 ML is shown in more detail. The characteristic InAs signatures appear of equal size, exhibiting a width of about 0.7 nm in [1 1 0] direction and a length of about 1.1 nm in [$\bar{1}$ 1 0] direction. Their height exceeds the surrounding surface by about 0.09 nm. These values are very similar to the InAs signatures observed on the GaAs-c(4×4) reconstructed surface as shown before.

Magenta arrows ① mark the characteristic trench along the [$\bar{1}$ 1 0] direction that separates the InAs signatures. The distance between the trenches in Fig. 8.5 a is 1.2 nm and thus corresponding to a threefold periodicity of surface unit cells. Thus the trench clearly does not originate from the initial GaAs(0 0 1)-β2(2×4) reconstructed surface, as it is not compatible with a fourfold periodicity.

However, at the areas, where the InAs signatures are found in a much closer packaging, a fourfold periodicity can be observed. In Fig. 8.5 a the corresponding (2×4) surface unit cell is illustrated by dotted yellow rectangles ②. Evidently, each (2×4) surface unit cell may provide an adsorption site for one InAs signature, arranging in an alternating alignment from one half-side to the other half-side at neighboring surface unit cells in a pronounced zig-zag chain. Considering this as the closest possible arrangement of InAs signatures at all, the observed total number of InAs signatures on sample ⟦F⟧ corresponds to a coverage of about 32% of the (2×4) surface unit cells. For the second layer InAs a small additional coverage share of about 1.6% is observed.

These findings on sample ⟦F⟧ are supported by the findings on sample ⟦G⟧ as well. In the STM image in Fig. 8.5 b the corresponding surface is shown in detail. With the increased deposition time the total number of InAs related signatures has increased just as well, resulting a in very even distribution. The characteristic trench exhibits a similar threefold (n×3) periodicity as observed on sample ⟦F⟧ with an intermediate distance of about 1.2 nm (marked by magenta arrows ③). In the perpendicular [1 1 0] direction a typical signature distance of about 1.6 nm is observed, corresponding to a fourfold periodicity of surface unit cells here (marked by green arrows ④). Thus the typical alignment of the InAs signatures can be described by (4×3) surface unit cells considering in-line arrangement or by c(4×6) surface unit cells considering brick-line arrangement (both of them exemplarily marked by yellow boxes).

The much closer alignment corresponding to a (2×4) symmetry could not be clearly observed. However, comparing the number of observed signatures on sample ⟦G⟧ to the number of possible initial (2×4) surface unit cells, a surface coverage share of about 45% is yielded. The number of the signatures with much brighter contrast from the second InAs layer has increased as well, corresponding to an additional coverage of (2×4) surface unit cells of about 4.2%.

The surface characteristics observed on sample [H] are different. In the STM image in Fig. 8.6 the sample surface is shown in detail. There are InAs related objects on top of an underlying surface that is clearly not the initial GaAs(0 0 1)-$\beta 2(2\times 4)$ reconstructed surface. There is a characteristic trench along the [$\bar{1}$ 1 0] direction that exhibits a threefold periodicity in [1 1 0] direction, similar to the characteristic trench on both samples before (marked with magenta arrows). Between the trenches the surface exhibits signatures of slightly brighter contrast that are aligned in a typical fourfold periodicity along [$\bar{1}$ 1 0] direction (marked with green arrows). Both findings may suggest a (4×3) or $c(4\times 6)$ reconstruction of the underlying surface.

The InAs objects on top cover about 31% of the underlying surface. They appear clearly different from the InAs signatures on the two samples discussed before. A typical alignment in zig-zag chains along the trench direction of the underlying surface is observed, exhibiting a twofold periodicity between the alternating objects, as exemplarily marked with blue arrows in Fig. 8.6. In review to the findings during growth on the GaAs(0 0 1)-$c(4\times 4)$ surface discussed in the previous Chapter 7, a (2×4) surface unit cell arrangement may be suggested for the InAs islands observed on top of the $(4\times 3)/c(4\times 6)$ reconstructed surface. The corresponding surface unit cells for the underlying (4×3) reconstruction and the (2×4) reconstructed InAs objects are exemplarily marked by yellow boxes in Fig. 8.6.

8.3.3. Discussion

Comparing the findings on InAs thin film growth on the GaAs(0 0 1)-$c(4\times 4)$ surface discussed in the previous chapter to the growth results on the GaAs(0 0 1)-$\beta 2(2\times 4)$ surface here, the strong similarities become evident. Thus the following discussion will be based on aspects already discussed in Chapter 7.

Samples [F] and [G] refer to the early deposition stages of InAs thin films on the GaAs(0 0 1)-$\beta 2(2\times 4)$ surface. The occasionally observed closed packaging of the InAs related signatures on sample [F] at a deposited InAs amount of about 0.31 ML exhibited an alternating alignment of (2×4) surface unit cells (cf. Fig. 8.5 a). Thus it is assumed that the InAs is first adsorbed on preferential adsorption sites of the initial GaAs(0 0 1)-$\beta 2(2\times 4)$ surface. Probably similar to the GaAs adsorption sites during homo-epitaxy [194], these adsorption sites may be at the top As dimers or the As trench dimers, as illustrated in Fig. 8.7. Strain issues caused by the slightly larger In atoms then may induce the observed alternating alignment of the adsorption sites at neighboring (2×4) surface unit cells. However, this would result in a height difference among the neighboring InAs signatures observed in the close (2×4) commensurate packaging, due to the structural difference of the two assumed possible adsorption sites in Fig. 8.7. This could yet not be observed. Unfortunately, there is no further STM data of material depositions less than 0.31 ML InAs to verify a specific growth model beginning from the early growth stages.

8.3. InAs thin films on GaAs(0 0 1)-β2(2×4)

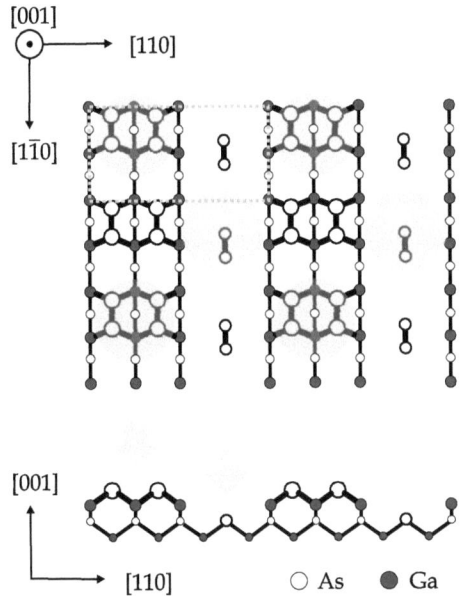

Figure 8.7: Structural model of the GaAs(0 0 1)-β2(2×4) reconstructed surface suggesting probable adsorption sites for InAs. Following the discussion in Ref. [194] for the adsorption of further GaAs during homoepitaxy, the top dimer sites and the trench dimer sites are energetically most favorable. These sites thus may also be attractive for InAs adsorption. The slightly larger In atoms then may cause an alternating alignment within adjacent surface unit cells to reduce strain effects. The resulting zig-zag alignment is illustrated by yellow circles corresponding to the suggested InAs adsorption sites.

As the close (2×4) arrangement of the InAs signatures was not observed at higher InAs depositions, such as 0.51 ML on sample [G], strain issues of the larger In atoms may possibly trigger another more favorable structural alignment. The vast majority of the observed InAs signatures was found to be aligned along a characteristic trench in [$\bar{1}$ 1 0] direction. The occurrence of this trench exhibits a threefold periodicity in the [1 1 0] direction. The very same phenomenon was observed on sample [C] at 0.56 ML InAs deposited on the GaAs(0 0 1)-c(4×4) surface. Thus the same mechanism that leads to the formation of such a structure may be assumed here.

During growth the highly mobile In atoms constantly switch between available surface adsorption sites and some may yet be chemisorbed. While during the InAs deposition onto the GaAs(0 0 1)-c(4×4) surface at $T_S = 460\,°C$ the desorption rate of In atoms is negligible, such effects become significant during the InAs deposition onto the GaAs(0 0 1)-β2(2×4) surface at $T_S = 530\,°C$, where the In desorption rate may exceed $0.0065\,\text{ML/s}$. Considering an average deposition rate of $0.010\,\text{ML/s}$, a large fraction of In atoms is desorbed again during growth. Moreover, an additional amount of 0.2 ML is desorbed during the 30 s growth interruption before quenching. These effects lead to a significant increase in the growth time τ for yielding the same remaining InAs deposition amounts after growth at $T_S = 530\,°C$ as compared to growth at $T_S = 460\,°C$. With the population of more adsorption sites and considering the high temperature level of $T_S = 530\,°C$ the increased probability of interchange processes with surface atoms may enhance the dissolution of the initial GaAs-β2(2×4) surface reconstruction during growth. Furthermore, the large τ may be responsible for the occurrence of this phenomena even at the lower InAs coverage of 0.31 ML on sample

[F] while such a dissolution of the underlying surface order could not be clearly observed at a comparable InAs coverage of 0.30 ML on sample [B] on the initial GaAs-c(4×4) reconstructed surface, but only at the slightly higher coverage of 0.56 ML of InAs with a slightly larger τ on sample [C].

Following the discussion in Chapter 7 assuming an InAs content of about eight In atoms per observed signature and considering the additional InAs occasionally observed on top of the signatures, the total coverages on samples [F] and [G] are estimated to an InAs content of 0.33 ML and 0.48 ML, respectively, in good agreement with the nominal deposition values (see Tab. 8.1).

As the periodicity of the signatures seems not to be commensurate to the initial surface reconstruction it is assumed that — similar to the suggestions in the previous discussion on the GaAs(0 0 1)-c(4×4) growth surface — it may be energetically more favorable for the deposited InAs to reorder into a different configuration. According to the findings in both growth regimes, this favorable configuration is an (n×3) reconstruction with a characteristic trench along the [$\bar{1}$ 1 0] direction.

With a further increase of the InAs coverage to about 0.92 ML of InAs, the findings on sample [H] appear very similar to the findings on sample [D] with about 0.73 ML of InAs deposited in the lower temperature growth regime $T_S = 460\,°C$ on the GaAs(0 0 1)-c(4×4) reconstructed surface. The detailed discussion of the findings on sample [D] revealed that a (4×3) reconstructed $In_{2/3}Ga_{1/3}As$ layer is formed during growth, when a sufficient amount of InAs is deposited. On top of this layer (2×4) reconstructed InAs islands evolve, forming mostly zig-zag chains due to the alternating sequence of $\alpha 2$ and $\alpha 2$-m configured surface unit cells.

The observations on sample [H] allow the assumption of a similar process here. The (4×3) reconstruction of the underlying surface is evident, as well as the (2×4) zig-zag alignment of the InAs islands on top. Unfortunately the atomic resolution of the presented data does not allow a more detailed analysis of the atomic structure, such as the one presented in Chapt. 7.

However, using the assumed structural model to calculate the InAs amount that was deposited on the surface, the (2×4) reconstructed islands cover about 31% of the surface corresponding to about 0.23 ML InAs content and about 0.67 ML are required for the underlying $In_{2/3}Ga_{1/3}As$-(4×3) reconstructed layer. The total InAs content of the WL on sample [H] then amounts to about 0.90 ML, again agreeing nicely with the nominal value (see Tab. 8.1).

Using these structural models, the assumption, that the initial InAs signatures observed in the early growth stages on samples [F] and [G], as well as on samples [A], [B], and [C], indeed contain about eight In atoms in average can be further supported. As suggested in the structural discussion, the first layer of InAs is expected to transform into an $In_{2/3}Ga_{1/3}As$-(4×3) reconstructed layer at a sufficient InAs amount, probably very close to the 0.67 ML of InAs necessary for a full coverage. However, considering the partly developed first layer as a pre-stage of this $In_{2/3}Ga_{1/3}As$-(4×3) layer, the number of In atoms here then must be sufficient to meet this amount of 0.67 ML of InAs when the surface is fully covered by the signatures. As the STM results show, a configuration of the signatures commensurate to a (4×3) periodicity is favored for InAs coverages close to 0.67 ML, and thus a full coverage of signatures corresponds to an average surface coverage of 67% with respect to the initial $\beta 2(2\times 4)/c(4\times 4)$ surface unit cells, if a pure InAs content is assumed for the signatures. These suggestions further imply, that each signature may be assigned to one (4×3) surface unit cell after the surface reconfiguration. Yet it is known from the structural discussion, that each (4×3) surface unit cell contains eight In atoms and four Ga atoms. This is in nice agreement with the assumed eight In atoms per InAs signature.

8.3.4. Summary

In principle, the evolution of InAs thin films grown on the GaAs(0 0 1)-$\beta 2(2\times 4)$ surface follows the same three stage growth model as discussed in Sects. 7.3–7.4 for InAs thin films grown on the GaAs(0 0 1)-c(4×4) surface.

In the first growth stage the InAs is adsorbed at designated surface sites of the underlying initial surface reconstruction. During growth the In atoms are highly mobile, thus the adsorption is mainly triggered by the rapid quenching. However, as the lowest amount of InAs on the investigated samples already was 0.31 ML, only fragmental areas of an alignment of the InAs signatures corresponding to the initial $\beta 2(2\times 4)$ reconstructed surface could be observed. This data unfortunately did not suffice for a detailed structural discussion of these signatures.

In the second growth stage the deposited InAs at the surface leads to the transformation into an $In_{2/3}Ga_{1/3}As$-(4×3) reconstructed ML. Pre-stages of this transition could be observed already at a deposited amount of about 0.31 ML and more clearly at a deposited amount of about 0.51 ML of InAs. The InAs adsorption at the surface is assumed to induce a dissolution of the underlying initial surface reconstruction. During rapid quenching an (n×3) reconstructed surface configuration is preferred, characterized by trenches along the [$\bar{1}$ 1 0] direction. The $In_{2/3}Ga_{1/3}As$-(4×3) reconstructed surface finally is observed after depositing an InAs amount exceeding 0.67 ML.

Considered as the third growth stage, further deposited InAs adsorbs on top of the $In_{2/3}Ga_{1/3}As$-(4×3) reconstructed surface in a (2×4) unit cell alignment. The mostly observed alignment is in zig-zag chains along the [$\bar{1}$ 1 0] direction, which corresponds to an alternating sequence of InAs-$\alpha 2(2\times 4)/\alpha 2(2\times 4)$-m reconstructed surface unit cells. Larger InAs objects then likewise correspond to InAs-$\beta 2(2\times 4)$ reconstructed surface unit cells.

The effectively grown InAs amounts were much lower than during growth on the GaAs-c(4×4) surface because of the significant desorption rate of InAs at the higher temperature level $T_S = 530\,°C$ for growth on the GaAs-$\beta 2(2\times 4)$ surface. As a consequence, the critical thickness for QD growth could not be achieved and thus QDs were not grown during these experiments.

9. Conclusion

In the present work, the evolution of the InAs WL grown by MBE on the GaAs(0 0 1) surface was studied using RHEED and STM. In contrast to detailed studies on the InAs/GaAs QD formation, the formation and structure of the corresponding WL so far remained largely uncertain. The investigations in this work focused on the detailed evolutionary pathway of InAs thin films in both typical growth regimes, on the GaAs(0 0 1)-c(4×4) at a substrate temperature $T_S = 460\,°C$ and the GaAs-$\beta2(2\times4)$ surface at a substrate temperature $T_S = 530\,°C$, starting from InAs submonolayer coverages to InAs QD formation. Based on the analysis of atomically resolved STM images, a principal three stage growth model for the InAs WL could be developed.

During the first growth stage depositing very small amounts of InAs, e.g. 0.09 ML, onto the GaAs(0 0 1)-c(4×4) reconstructed substrate surface, characteristic InAs signatures were observed, each characterized by an average In content of eight In atoms. On the c(4×4) reconstructed surface, these signatures assemble only at the hollow sites, preferentially near surface defect sites and domain boundaries. Supported by DFT calculations, the adsorption of In atoms at the hollow sites is considered energetically favorable. Furthermore, a strain discussion on the c(4×4) surface showed that an additional incorporation of In atoms at the hollow sites will not significantly induce further strain. Based on these results, a model for the incorporation of the first six In atoms at each surface unit cell was derived, yet none of the structural models fulfilled the ECR. A model of the final signature structure could not be clarified, due to a lack of resolution in the local STM data. This lack of local resolution is ascribed to electronic effects from excess surface electrons or a possible quasi-metallic character of the signatures themselves. There was no sufficient data on the $\beta2(2\times4)$ surface for very low amounts of deposited InAs to verify such a structural model for this growth regime. However, at a deposited amount of 0.31 ML of InAs a (2×4) periodic alignment of the signatures was, rarely though, observed.

With the number of signatures increasing proportionally with increased amounts of deposited InAs, their typical size and appearance remained unchanged. After the deposition of 0.56 ML of InAs onto the GaAs(0 0 1)-c(4×4) surface, the initial surface structure could not be verified any more. At this coverage, the InAs signatures were observed evenly distributed across the surface, but rather aligned in an (n×3) periodic alignment, incommensurate to the prior (4×4) periodicity of the initial surface. In principle, the same phenomena was observed on the GaAs(0 0 1)-$\beta2(2\times4)$ for deposited amounts of 0.31 ML and 0.51 ML of InAs.

From these findings it is concluded for both growth regimes, that the highly mobile In atoms very likely move between neighboring adsorption sites during growth. Some In atoms chemisorb at favorable surface sites, attracting the remaining In atoms to condense at these sites during the rapid quenching, but do not have to chemisorb necessarily. This leads to the formation of the InAs signatures. With increasing amounts of InAs during growth and the extended growth time at the substrate temperature, interchange processes with surface atoms become more likely, leading to a disintegration of the underlying initial surface structure. At the much higher substrate temperature in the $\beta 2(2\times 4)$ growth regime at $T_S = 530\,°C$, the desorption rate of surface In atoms is significantly increased to about $2/3$ of the average deposition rate. The resulting larger growth time is assumed to support the disintegration of the initial surface structure, which explains why an initial (2×4) commensurate alignment of the InAs signatures was only rarely observed here. As a result, during quenching, the disintegrated underlying surface reconstruction is inadequate to induce a structure commensurate to its initial configuration. The observed $(n\times 3)$ surface reconstruction evidently is more favorable.

In the second growth stage, at a deposited InAs amount around 0.67 ML the surface structure transforms into a very stable (4×3) reconstructed $In_{2/3}Ga_{1/3}As$ ML, for which a detailed structural model was derived. This transformation was clearly observed in both growth regimes. The $In_{2/3}Ga_{1/3}As(0\,0\,1)$-$(4\times 3)$ surface is characterized by three As top dimers and a stable trench with further As trench dimers. Occasionally inserted $\eta(6\times 3)$ and $\eta(2\times 3)$ surface unit cells shift the alignment of the triple dimer blocks between in-line and brick-lined. In the brick-lined alignment, the common surface structure is described by a $c(4\times 6)$ unit cell. The $In_{2/3}Ga_{1/3}$ stoichiometry of the InGaAs layer was shown to efficiently reduce the compressive strain resulting from the incorporation of slightly larger In atoms at Ga atom sites, which is further reduced by the formation of As dimers at the stable trench.

In the third growth stage, further deposited InAs arranges in small 2D islands on top of the first $In_{2/3}Ga_{1/3}As$ layer. Three slightly different configurations were observed for these islands, all of them corresponding to a (2×4) periodicity, and the corresponding structural models were presented. The STM data could be further verified by corresponding DFT calculations of the STM contrast. Mostly an alternating alignment of InAs-$\alpha 2(2\times 4)/\alpha 2(2\times 4)$-m surface unit cells was observed, resulting in zig-zag chains along the $[\bar{1}\,1\,0]$ surface direction. This structure accumulates unfavorable amounts of compressive strain that, however, cannot be as efficiently compensated as within the previous structures. Even though the more As-rich InAs-$\beta 2(2\times 4)$ structure accumulates less strain, it is less often observed.

Upon further deposition of InAs, the surface coverage by this second (2×4) reconstructed InAs layer increases. As its structure is incommensurate to the (4×3) reconstruction of the underlying $In_{2/3}Ga_{1/3}As$ ML, the disintegration of the underlying As triple dimers inevitably must occur during overgrowth, further increasing the unfavorable compressive strain situation. Thus, upon the completion of the second InAs layer, the accumulated strain energy leads to the 2D→3D growth transition and small InAs QD precursors evolve. At this

9. Conclusion

point, the completed two-layer WL contains 1.42 ML of InAs in total, which is considered the actual critical thickness. The QD precursors act as an attractive site for the accumulation of further InAs, so that QDs evolve finally. Moreover, it is observed that some InAs material from the WL must have relocated into the QDs, presumably due to strain issues. The QDs then undergo ripening processes by accumulating further deposited InAs material as well as consuming it from the WL. The appearing disagreement between the critical thickness of 1.42 ML of InAs derived here and prior reports, mostly relying on diffraction methods like RHEED, assuming a higher critical thickness of about 1.6 ML of InAs can be explained by considering the evolution of the QDs as a time dependent process and the sensitivity of RHEED. For a sufficient contribution of QDs to the RHEED diffraction pattern, a significant number of QDs must have evolved. Thus the initial 2D→3D of the InAs precursors and early QDs may very likely not be observed yet, resulting in a slight overestimation of the actual critical thickness.

Upon deposition of a total amount of 1.6 ML of InAs onto the initial GaAs(0 0 1)-c(4×4) surface, a typical QD density of about $8.3 \cdot 10^{10} \, cm^{-2}$ was observed. The WL was found partly uncovered by a share of about 15%, corresponding to a missing share of 0.11 ML of InAs material, that probably accumulated in the QDs. Due to the high InAs desorption rate in the higher temperature growth regime on the GaAs(0 0 1)-β2(2×4) surface the critical thickness for QD growth could not be achieved and thus QD could not be grown on that surface.

The findings in this thesis have led to a detailed pathway of the InAs WL evolution on the GaAs(0 0 1) surface. However, some details, in particular on the structure of the initial InAs signatures, on the mechanisms at the transition to the (4×3) reconstructed In$_{2/3}$Ga$_{1/3}$As ML, and on the QD growth on the GaAs-β2(2×4) surface, are still missing. To avoid the significant desorption of InAs during growth, a substrate temperature of about 500–510 °C seems more reasonable for growth studies on the GaAs-β2(2×4) surface. As this temperature is close to the β2(2×4)/c(4×4) surface transition, reliable RHEED observations during growth are indispensable. Thus further experiments are important to further complete the presented evolutionary pathway of InAs thin films on the GaAs(0 0 1) surface.

Appendix

A. Data of the beam equivalent pressure (BEP)

The averaged BEP data of the constituent effusion cells used in the investigations of this work is presented in the following. This data is used to confirm the proper operation of the K-cells on a regular basis. The BEP is displayed as a function of the cell temperature T. The data points are derived from an ion gauge placed at the sample position in the MBE setup, the red curve is derived from a data fit following a Boltzmann-like distribution of the form $BEP = p_{rg} + C_\alpha \, exp(\frac{-C_\beta}{T})$, with constant values C_α and C_β and a constant contribution p_{rg} from the remaining residual gas pressure.

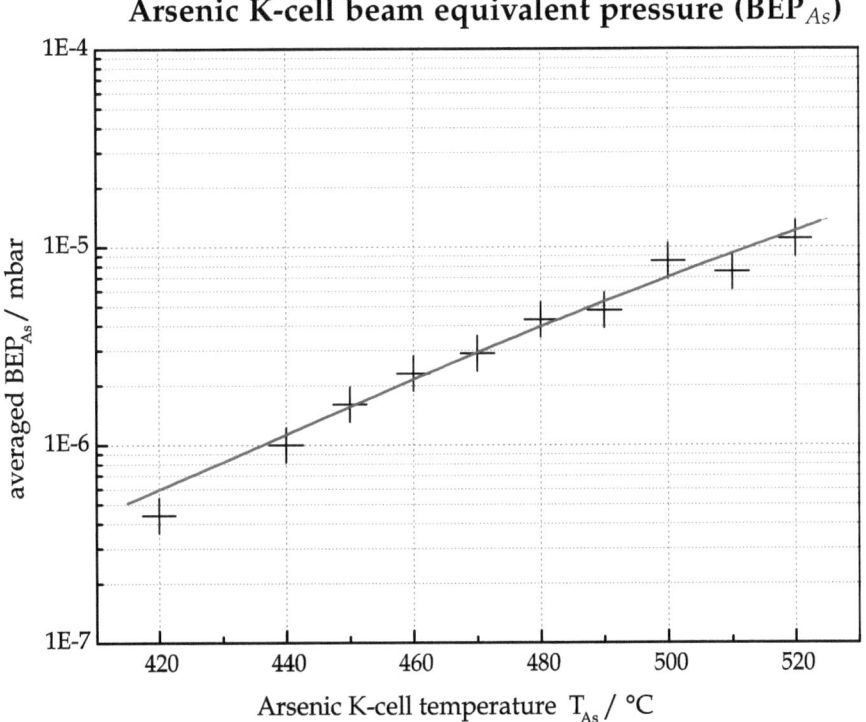

Figure A.1: *Averaged BEP data of the As K-cell for the effusion of As_2 molecules at a constant cracker temperature $T_{Cr} = 985°C$ in correlation to the cell temperature T_{As}.*

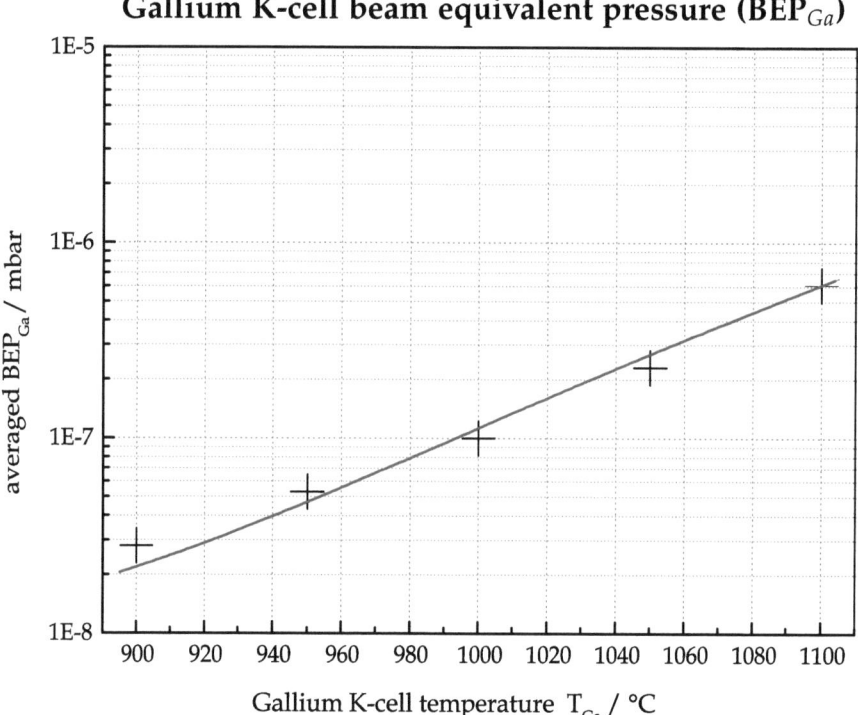

Figure A.2: Averaged BEP data of the Ga K-cell for the effusion of Ga atoms in correlation to the cell temperature T_{Ga}.

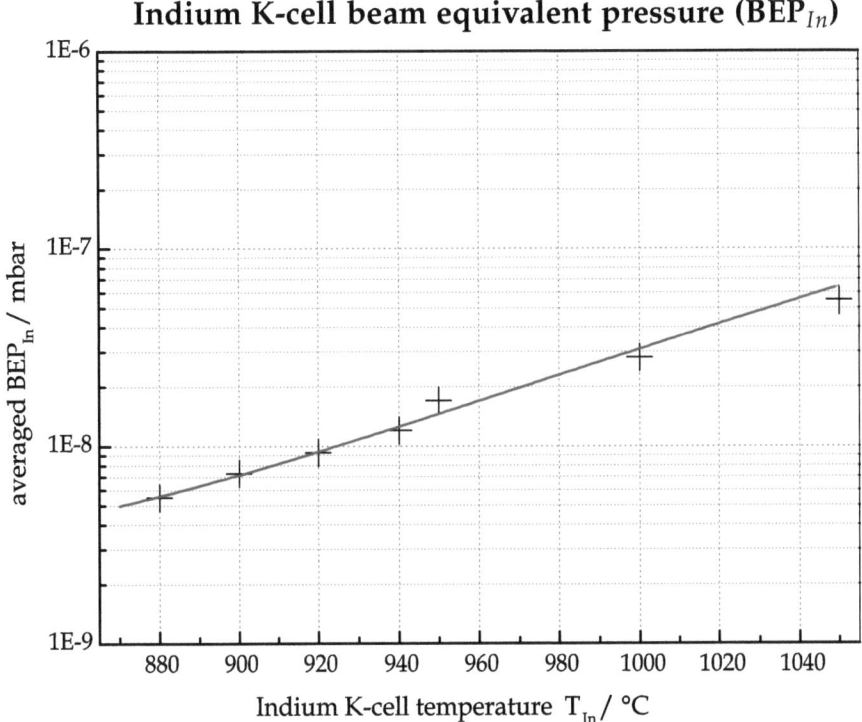

Figure A.3: *Averaged BEP data of the In K-cell for the effusion of In atoms in correlation to the cell temperature T_{In}.*

B. Determination of the deposited InAs coverage

Due to problems with the shutter operation, the In flux during InAs growth could only be controlled by the cell temperature T_{In} of the In K-cell. Growing under As-rich conditions, the In flux is the determining parameter for the amount of adsorbed InAs at the surface. Following the definition of the material flux j in Chapt. 3, in a static setup the effective In flux j_{In} from the source to the sample basically is determined by the number of impinging particles per time (cf. Eqs. 3.5–3.8). The rate of effectively adsorbed In atoms is described by its corresponding sticking coefficient α_{In} (cf. Eq. 3.9). Finally, the number of impinging particles per time is correlated to the BEP and cell temperature T by the Knudsen effusion equation (Eq. 3.4). Consequently, the rate of adsorbed In atoms during growth applies to

$$\frac{dN_{In}}{dt} \propto BEP_{In}(T_{In}) \cdot \sqrt{\frac{1}{T_{In}}}. \tag{B.1}$$

Integrating $\frac{dN_{In}}{dt}$ over the deposition time τ then yields the total number of adsorbed In atoms after growth which can be directly converted into the InAs coverage thickness.

The correlation between the BEP_{In} and the cell temperature T_{In} is described in Fig. A.3. However, due to the inoperability of the shutter of the K-cell and thereby the impossibility to block the effused In beam during the heating of the In K-cell until the desired operation temperature is achieved, there is a temperature variation of T_{In} during the time of operation (Fig. B.1). As a consequence the BEP_{In} is also time dependent, following the temperature variation

$$T_{In} = T_{aim} - C_\alpha \exp(-C_\beta \cdot t_{In}) \tag{B.2}$$

with constant values C_α and C_β, assuming exponential saturation towards T_{aim}, the aimed cell temperature.

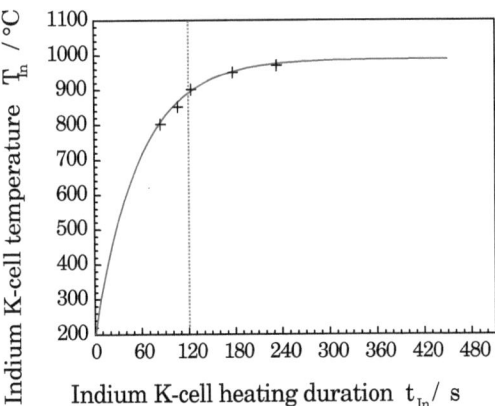

Figure B.1: Time dependent temperature gradient of T_{In} during the heating of the In K-cell to an appropriate effusion temperature. Data points are derived from the K-cell thermocouple. The red curve illustrates a data fit according to Eq. B.2. The dotted line at $t_{In} = 120s$ marks the starting point of the considered deposition time τ, as the In growth rate before the corresponding cell temperature was considered to be negligible.

It may be assumed that there is no considerable In effusion from the In K-cell during heating as long as the BEP_{In} is in the order of the residual gas pressure p_{rg}. An appropriate effusion cell temperature is reached when the BEP_{In} clearly exceeds the value of the residual gas pressure ($BEP_{In} \gg p_{rg}$), which is estimated from the data in Figs. A.3 and B.1 at $T_{In} >$ 880 °C or $t_{In} = 120$ s, respectively. As a consequence, InAs growth is considered to start at $t_{In} = \tau_0 = 120$ s and to endure for a deposition time τ defined as $\tau = t_{In} - \tau_0$. After τ has elapsed, the heating current of the In K-cell is switched off, resulting in an immediate decrease of BEP_{In} below the residual gas pressure.

Experimental data

		— growth on the GaAs(0 0 1)-c(4×4) surface —		
		sample A	sample B	sample C
deposition time τ		20 s	52 s	80 s
1st layer	coverage of c(4×4) unit cells*	10±3%	30±4%	50±5%
estimated total InAs content derived from the STM data		0.10±0.03 ML	0.30±0.04 ML	0.50±0.05 ML
		sample D	sample E	
deposition time τ		95 s	170 s	
1st layer	coverage by (4×3) unit cells**	100% (by def.)	100% (by def.)	
	estimated InAs content	0.67 ML (by def.)	0.67 ML (by def.)	
2nd layer	coverage by (2×4) unit cells*	15±5%	85±5%	
	estimated InAs content	0.11±0.04 ML	0.64±0.04 ML	
QDs and precursors	QD density	-	$\approx 8.3 \cdot 10^{10}$ cm^{-2}	
	precursor density	-	$\approx 1.6 \cdot 10^{10}$ cm^{-2}	
	estimated InAs content	-	0.29±0.05 ML	
estimated total InAs content derived from the STM data		0.78±0.04 ML	1.60±0.09 ML	

		— growth on the GaAs(0 0 1)-β2(2×4) surface —		
		sample F	sample G	sample H
deposition time τ		170 s	200 s	260 s
1st layer	coverage of (2×4) unit cells*	32±5%	45±5%	-
	coverage by (4×3) unit cells**	-	-	100% (by def.)
	estimated InAs content	0.32±0.05 ML	0.45±0.05 ML	0.67 ML (by def.)
2nd layer	coverage by (2×4) unit cells*	1.6±0.5%	4.2±0.5%	31±5%
	estimated InAs content	0.01±0.01 ML	0.03±0.01 ML	0.23±0.04 ML
estimated total InAs content derived from the STM data		0.33±0.05 ML	0.48±0.06 ML	0.90±0.04 ML

Table B.1: *InAs coverage data of the investigated samples as derived from the STM data and the structural discussion. The InAs coverage values were estimated using the structural models of the respective growth stage, as discussed in detail in Chapters 7 and 8.*
* *The surface areas of the c(4×4) and the (2×4) surface unit cells each correspond to $A = 1.28$ nm^2.*
** *The surface area of one (4×3) surface unit cell corresponds to $A = 1.92$ nm^2.*

Following Eqs. B.1 and B.2, the actual InAs growth rate and thereby the resulting InAs coverage of the sample surface was numerically integrated using the fit curves in Figs. A.3 and B.1, resulting in the InAs growth function shown in Fig. B.2. The respective proportionality constant was adjusted to minimize the tolerance between the integrated values (blue curve) and the actual coverage data (blue data points), derived from the STM observations (Tab. B.1).

The blue curve applies to the effective InAs coverage in the lower temperature growth regime on the GaAs(0 0 1)-c(4×4) reconstructed surface at $T_S = 460\,°C$. The red curve applies to the effective InAs coverage in the higher temperature growth regime on the GaAs(0 0 1)-$\beta 2(2 \times 4)$ reconstructed surface at $T_S = 530\,°C$. The red curve is derived by considering an additional InAs desorption rate of 0.0067 ML/s and an additional amount of 0.20 ML InAs desorbed during the 30 s growth interruption after deposition.

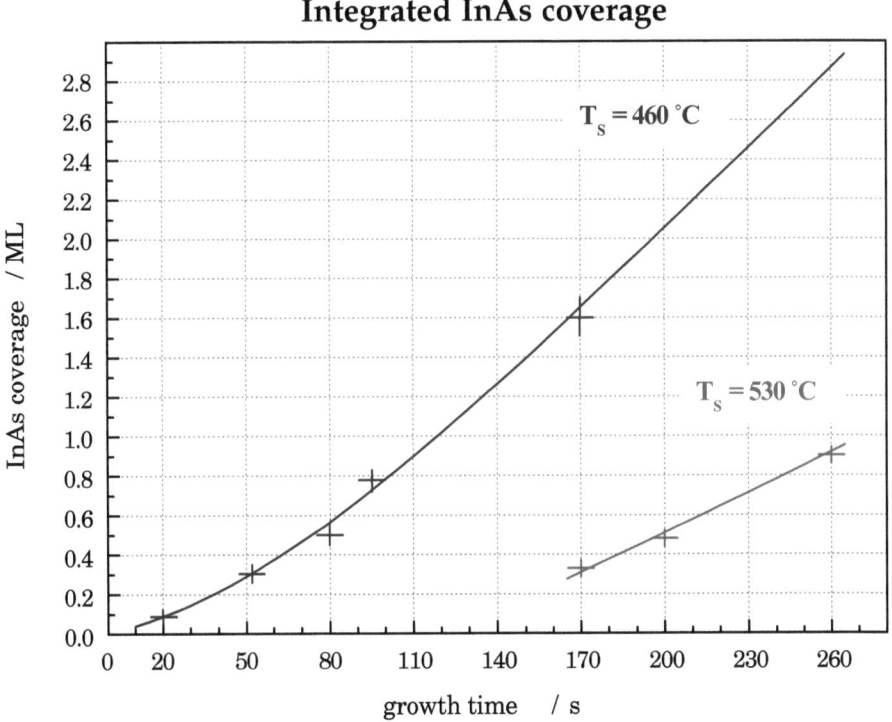

Figure B.2: *InAs growth function derived from the numerical integration of the effective deposition rate during the growth time τ. Data points represent the respective experimental data from Tab. B.1.*

C. Sample heating data

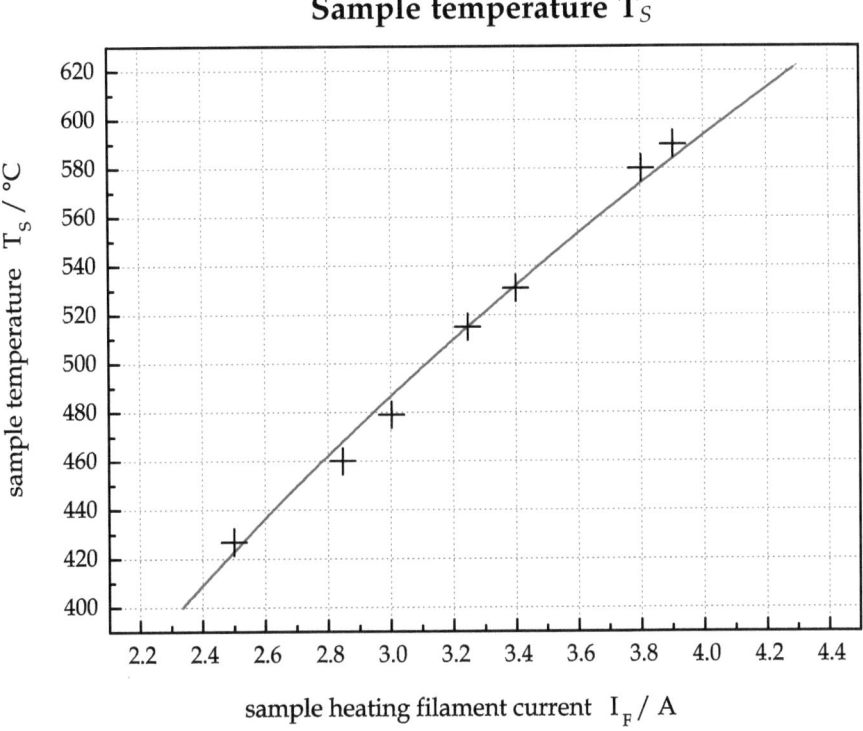

Figure C.1: Sample temperature in correlation to the filament current of the sample heating filament. Data points are derived from an optical pyrometer facing the sample surface via a chamber window. The red curve is derived from a data fit of the form $I_F^2 \propto \alpha T + \beta T^4$, with constant values α and β. In the relation the contributions of the thermal conduction (following Fourier's law) and the thermal radiation (following the Stefan-Boltzmann law) are considered.

List of Abbreviations

2D	two-dimensional
3D	three-dimensional
BEP	beam equivalent pressure
CB	conduction band
db	dangling bond(s)
ECR	electron counting rule
GaAs	gallium arsenide
IBA	ion bombardment and annealing
InAs	indium arsenide
K-cell	Knudsen cell
LDS	low-dimensional semiconductor structures
MBE	molecular beam epitaxy
ML	monolayer
QD	quantum dot
RHEED	reflection high energy electron diffraction
SK growth	Stranski-Krastanow growth
STM	scanning tunneling microscopy
UHV	ultra high vacuum
VB	valence band
WL	wetting layer

Bibliography

[1] W.F. Brinkman and D.V. Lang, Rev. Mod. Phys. **71**, S480 (1999).

[2] J. Bardeen and W.H. Brattain, Phys. Rev. **74**, 230 (1948).

[3] M. Grundmann, *The Physics of Semiconductors*, Springer, Berlin, 2006.

[4] R.W. Keyes, IBM J. Res. Dev. **32**, 24 (1988).

[5] P.Y. Yu and M. Cardona, *Fundamentals of Semiconductors. Physics and Material Properties*, Springer, Berlin, 1996.

[6] M.A. Kastner, Phys. Today **46**, 24 (1993).

[7] M.A. Reed, Sci. Am. **268**, 118 (1993).

[8] D. Bimberg (Ed.), *Semiconductor Nanostructures*, Springer, Berlin, 2008.

[9] Y. Arakawa and H. Sakaki, Appl. Phys. Lett. **40**, 939 (1982).

[10] N. Kirstaedter, N.N. Ledentsov, M. Grundmann, D. Bimberg, V.M. Ustinov, S.S. Ruvimov, M.V. Maximov, P.S. Kop'ev, Zh.I. Alferov, U. Richter, P. Werner, U. Gösele, and J. Heydenreich, Electron. Lett. **30**, 1416 (1994).

[11] N.N. Ledentsov, V.M. Ustinov, A.Y. Egorov, A.E. Zhukov, M.V. Maksimov, I.G. Tabatadze, and P.S. Kopev, Semiconductors **28**, 832 (1994).

[12] S. Fafard, Z.R. Wasilewski, C.N. Allen, K. Hinzer, J.P. McCaffrey, and Y. Feng, Appl. Phys. Lett. **75**, 986 (1999).

[13] E. Mairiaux, L. Desplanque, X. Wallart, and M. Zaknoune, IEEE Electron. Dev. Lett. **31**, 299 (2010).

[14] M.A. Rowe, G.M. Salley, E.J. Gansen, S.M. Etzel, S.W. Nam, and R.P. Mirin, J. Appl. Phys. **107** (2010).

[15] H. Kosaka, T. Inagaki, Y. Rikitake, H. Imamura, Y. Mitsumori, and K. Edamatsu, Nature **457**, 702 (2009).

[16] E. Yablonovitch, H.W. Jiang, H. Kosaka, H.D. Robinson, D.S. Rao, and T. Szkopek, Proc. IEEE **91**, 761 (2003).

[17] L. Goldstein, F. Glas, J.Y. Marzin, M.N. Charasse, and G. Leroux, Appl. Phys. Lett. **47**, 1099 (1985).

[18] P. Schittenhelm, C. Engel, F. Findeis, G. Abstreiter, A.A. Darhuber, G. Bauer, A.O. Kosogov, and P. Werner, J. Vac. Sci. Technol. B **16**, 1575 (1998).

[19] N.H. Bonadeo, J. Erland, D. Gammon, D. Park, D.S. Katzer, and D.G. Steel, Science **282**, 1473 (1998).

[20] V.M. Ustinov, E.R. Weber, S. Ruvimov, Z. Liliental-Weber, A.E. Zhukov, A.Yu. Egorov, A.R. Kovsh, A.F. Tsatsul'nikov, and P.S. Kop'ev, Appl. Phys. Lett. **72**, 362 (1998).

[21] L. Vegard, Z. Phys. **5**, 17 (1921).

[22] A.R. Denton and N.W. Ashcroft, Phys. Rev. A **43**, 3161 (1991).

[23] R. Driad, Z.H. Lu, S. Charbonneau, W.R. McKinnon, S. Laframboise, P.J. Poole, and S.P. McAlister, Appl. Phys. Lett. **73**, 665 (1998).

[24] T. Lundstrom, W. Schoenfeld, H. Lee, and P.M. Petroff, Science **286**, 2312 (1999).

[25] C.-S. Lee, F.-Y. Chang, D.-S. Liu, and H.-H. Lin, Jpn. J. Appl. Phys. **45**, 6271 (2006).

[26] H. Shimizu, S. Saravanan, J. Yoshida, S. Ibe, and N. Yokouchi, Jpn. J. Appl. Phys. **46**, 638 (2007).

[27] S. Pyun and W. Jeong, J. Kor. Phys. Soc. **51**, 2010 (2007).

[28] L.H. Li, M. Rossetti, G. Patriarche, and A. Fiore, J. Crys. Growth **301-302**, 959 (2007).

[29] G. Trevisi, L. Seravalli, P. Frigeri, and S. Franchi, Nanotechnology **20** (2009).

[30] T. D. Germann, A. Strittmatter, J. Pohl, U.W. Pohl, D. Bimberg, J. Rautiainen, M. Guina, and O.G. Okhotnikov, Appl. Phys. Lett. **92** (2008).

[31] D.Z.-Y. Ting, S.V. Bandara, S.D. Gunapala, J.M. Mumolo, S.A. Keo, C.J. Hill, J.K. Liu, E.R. Blazejewski, Sir B. Rafol, and Y.-C. Chang, Appl. Phys. Lett. **94** (2009).

[32] F. Hopfer, A. Mutig, G. Fiol, M. Kuntz, V.A. Shchukin, V.A. Haisler, T. Warming, E. Stock, S.S. Mikhrin, I.L. Krestnikov, D.A. Livshits, A.R. Kovsh, C. Bornholdt, A. Lenz, H. Eisele, M. Dähne, N.N. Ledentsov, and D. Bimberg, IEEE J. Sel. Top. Quant. Electron. **13**, 1302 (2007).

[33] N.N. Ledentsov, F. Hopfer, A. Mutig, V.A. Shchukin, A.V. Savel'ev, G. Fiol, M. Kuntz, V.A. Haisler, T. Warming, E. Stock, S.S. Mikhrin, A.R. Kovsh, C. Bornholdt, A. Lenz, H. Eisele, M. Dähne, N. D. Zakharov, P. Werner, and D. Bimberg, Proc. SPIE **6468**, 64681O (2007).

[34] P. Westbergh, J.S. Gustavsson, A. Haglund, A. Larsson, F. Hopfer, G. Fiol, D. Bimberg, and A. Joel, Electron. Lett. **45**, 366 (2009).

[35] D. Leonard, K. Pond, and P.M. Petroff, Phys. Rev. B **50**, 11687 (1994).

[36] B.A. Joyce, D.D. Vvedensky, G.R. Bell, J.G. Belk, M. Itoh, and T.S. Jones, Mater. Sci. Eng. B **67**, 7 (1999).

[37] L. G. Wang, P. Kratzer, N. Moll, and M. Scheffler, Phys. Rev. B **62**, 1897 (2000).

[38] J.B. Nah, S.H. Park, K.M. Kim, Y.J. Park, C.K. Hyon, and E.K. Kim, J. Kor. Phys. Soc. **39**, 132 (2001).

[39] J. Márquez, L. Geelhaar, and K. Jacobi, Appl. Phys. Lett. **78**, 2309 (2001).

[40] H. Eisele and K. Jacobi, Appl. Phys. Lett. **90**, 129902 (2007).

[41] A. Lenz, R. Timm, H. Eisele, Ch. Hennig, S.K. Becker, R.L. Sellin, U.W. Pohl, D. Bimberg, and M. Dähne, Appl. Phys. Lett. **81**, 5150 (2002).

[42] H. Eisele, A. Lenz, R. Heitz, R. Timm, M. Dähne, Y. Temko, T. Suzuki, and K. Jacobi, J. Appl. Phys. **104**, 124301 (2008).

[43] J.H. Blokland, M. Bozkurt, J.M. Ulloa, D. Reuter, A.D. Wieck, P.M. Koenraad, P.C.M. Christianen, and J.C. Maan, Appl. Phys. Lett. **94**, 023107 (2009).

[44] T. Mattila, L.W. Wang, and A. Zunger, Phys. Rev. B **59**, 15270 (1999).

[45] H. Ibach and H. Lüth, *Festkörperphysik*, 6th edition, Springer, Berlin, 2002.

[46] Ch. Kittel, *Einführung in die Festkörperphysik*, 14th edition, Oldenbourg, München, 2005.

[47] W.A. Harrison, J. Vac. Sci. Technol. **16**, 1492 (1979).

[48] R.W.G. Wyckoff, *Crystal Structures*, 2nd Ed. Vol. 1, John Wiley & Sons, New York, 1965.

[49] C.B. Duke, Chem. Rev. **96**, 1237 (1996).

[50] B. Kübler, W. Ranke, and K. Jacobi, Surf. Sci. **92**, 519 (1980).

[51] R.M. Feenstra, J.A. Stroscio, J. Tersoff, and A.P. Fein, Phys. Rev. Lett. **58**, 1192 (1987).

[52] J.L.A. Alves, J. Hebenstreit, and M. Scheffler, Phys. Rev. B **44**, 6188 (1991).

[53] A.R. Lubinsky, C.B. Duke, B.W. Lee, and P. Mark, Phys. Rev. Lett. **36**, 1058 (1976).

[54] H. Eisele, *Cross-Sectional Scanning Tunneling Microscopy of InAs/GaAs Quantum Dots*, PhD thesis, Technische Universität Berlin, 2002.

[55] G.P. Srivastava, Appl. Surf. Sci. **252**, 7600 (2006).

[56] E. Penev, P. Kratzer, and M. Scheffler, Phys. Rev. Lett. **93**, 146102 (2004).

[57] N. Moll, A. Kley, E. Pehlke, and M. Scheffler, Phys. Rev. B **54**, 8844 (1996).

[58] J. Platen, C. Setzer, W. Ranke, and K. Jacobi, Appl. Surf. Sci. **123-124**, 43 (1998).

[59] J. Márquez, P. Kratzer, L. Geelhaar, K. Jacobi, and M. Scheffler, Phys. Rev. Lett. **86**, 115 (2001).

[60] L. Geelhaar, J. Márquez, P. Kratzer, and K. Jacobi, Phys. Rev. Lett. **86**, 3815 (2001).

[61] Y. Temko, L. Geelhaar, T. Suzuki, and K. Jacobi, Surf. Sci. **513**, 328 (2002).

[62] J.A. Appelbaum, G.A. Baraff, and D.R. Hamann, Phys. Rev. B **14**, 1623 (1976).

[63] M.D. Pashley, Phys. Rev. B **40**, 10481 (1989).

[64] L.J. Whitman, P.M. Thibado, S.C. Erwin, B.R. Bennett, and B.V. Shanabrook, Phys. Rev. Lett. **79**, 693 (1997).

[65] L. Däweritz and R. Hey, Surf. Sci. **236**, 15 (1990).

[66] A. Ohtake, P. Kocán, J. Nakamura, A. Natori, and N. Koguchi, Phys. Rev. Lett. **92**, 236105 (2004).

[67] F. Bastiman, A.G. Cullis, and M. Hopkinson, Surf. Sci. **603**, 2398 (2009).

[68] R. Heitz, T. R. Ramachandran, A. Kalburge, Q. Xie, I. Mukhametzhanov, P. Chen, and A. Madhukar, Phys. Rev. Lett. **78**, 4071 (1997).

[69] N.P. Kobayashi, T.R. Ramachandran, P. Chen, and A. Madhukar, Appl. Phys. Lett. **68**, 3299 (1996).

[70] E. Steimetz, J.-T. Zettler, F. Schienle, T. Trepk, T. Wethkamp, W. Richter, and I. Sieber, Appl. Surf. Sci. **107**, 203 (1996).

[71] J.G. Belk, J.L. Sudijono, D.M. Holmes, C.F. McConville, T.S. Jones, and B.A. Joyce, Surf. Sci. **365**, 735 (1996).

[72] O. Bute, Gh.V. Cimpoca, E. Placidi, F. Arciprete, F. Patella, M. Fanfoni, and A. Balzarotti, J. Optoelectron. & Adv. Mater. **10**, 74 (2008).

[73] A. Ohtake, J. Nakamura, S. Tsukamoto, N. Koguchi, and A. Natori, Phys. Rev. Lett. **89**, 206102 (2002).

[74] D.K. Biegelsen, R.D. Bringans, J.E. Northrup, and L.-E. Swartz, Phys. Rev. B **41**, 5701 (1990).

[75] D.K. Biegelsen, R.D. Bringans, J.E. Northrup, and L.-E. Swartz, Phys. Rev. B **42**, 3195 (1990).

[76] D.K. Biegelsen, R.D. Bringans, J.E. Northrup, and L.-E. Swartz, Phys. Rev. Lett. **65**, 452 (1990).

[77] M. Sauvage-Simkin, R. Pinchaux, J. Massies, P. Calverie, N. Jedrecy, J. Bonnet, and I. K. Robinson, Phys. Rev. Lett. **62**, 563 (1989).

[78] J.E. Northrup and S. Froyen, Phys. Rev. Lett. **71**, 2276 (1993).

[79] A. Ohtake and N. Koguchi, Appl. Phys. Lett. **83**, 5193 (2003).

[80] C. Hogan, E. Placidi, and R. Del Sole, Phys. Rev. B **71**, 041308 (2005).

[81] S. Kunsági-Máté, C. Schür, T. Marek, and H.P. Strunk, Phys. Rev. B **69**, 193301 (2004).

[82] L.H. Li, M. Rossetti, and A. Fiore, J. Cryst. Growth **278**, 680 (2005).

[83] S. Kamprachum, S. Thainoi, S. Kanjanachuchai, and S. Panyakeow, *Multi-Stacked InAs/GaAs Quantum Dot Structures and their Photovoltaic Characteristics*, Technical report, Chulalongkorn University, Bangkok, 2006.

[84] V.P. LaBella, H. Yang, D.W. Bullock, and P.M. Thibado, Phys. Rev. Lett. **83**, 2989 (1999).

[85] D.J. Chadi, J. Vac. Sci. Technol. A **5**, 834 (1987).

[86] T. Ohno, Phys. Rev. Lett. **70**, 631 (1993).

[87] J.E. Northrup and S. Froyen, Phys. Rev. B **50**, 2015 (1994).

[88] L.F. Lester, A. Stintz, H. Li, T.C. Newell, E.A. Pease, B.A. Fuchs, and K.J. Malloy, IEEE Photon. Technol. Lett. **11**, 931 (1999).

[89] A. Scherer and H.G. Craighead, Appl. Phys. Lett. **49**, 1284 (1986).

[90] A. Forchel, H. Leier, B.E. Maile, and R. Germann, Adv. Sol. St. Phys. **28**, 99 (1988).

[91] J. Tersoff, C. Teichert, and M.G. Lagally, Phys. Rev. Lett. **76**, 1675 (1996).

[92] M. Sugawara (Ed.), *Self-Assembled InGaAs/GaAs Quantum Dots*, Academic Press, London, 1999.

[93] F.C. Frank and J.H. van der Merwe, Proc. Roy. Soc. A **198**, 205 (1949).

[94] M. Volmer and A. Weber, Z. Phys. Chem. **119**, 227 (1926).

[95] V.A. Shchukin, N.N. Ledentsov, P.S. Kop'ev, and D. Bimberg, Phys. Rev. Lett. **75**, 2968 (1995).

[96] I.N. Stranski and L. Krastanow, Sitzungsbericht Akad. Wiss. Wien, Kl. IIB **146**, 797 (1938).

[97] D.J. Eaglesham and M. Cerullo, Phys. Rev. Lett. **64**, 1943 (1990).

[98] N. Carlsson, W. Seifert, A. Petersson, P. Castrillo, M.E. Pistol, and L. Samuelson, Appl. Phys. Lett. **65**, 3093 (1994).

[99] B. Daudin, F. Widmann, G. Feuillet, Y. Samson, M. Arlery, and J.L. Rouviére, Phys. Rev. B **56**, R7069 (1997).

[100] L. Müller-Kirsch, R. Heitz, U.W. Pohl, D. Bimberg, I. Häusler, H. Kirmse, and W. Neumann, Appl. Phys. Lett. **79**, 1027 (2001).

[101] J.M. Moison, F. Houzay, F. Barthe, L. Leprince, E. André, and O. Vatel, Appl. Phys. Lett. **64**, 196 (1994).

[102] T.J. Krzyzewski and T.S. Jones, J. Appl. Phys. **96**, 668 (2004).

[103] T. Kita, K. Tachikawa, H. Tango, K. Yamashita, and T. Nishino, Appl. Surf. Sci. **159–160**, 503 (2000).

[104] M. Sauvage-Simkin, Y. Garreau, R. Pinchaux, M.B. Véron, J.P. Landesman, and J. Nagle, Phys. Rev. Lett. **75**, 3485 (1995).

[105] Y. Garreau, K. Aïd, M. Sauvage-Simkin, R. Pinchaux, C.F. McConville, T.S. Jones, J.L. Sudijono, and E.S. Tok, Phys. Rev. B **58**, 16177 (1998).

[106] J. Mirecki Millunchick, A. Riposan, B.J. Dall, C. Pearson, and B.G. Orr, Surf. Sci. **550**, 1 (2004).

[107] A. Riposan, J. Mirecki Millunchick, and C. Pearson, J. Vac. Sci. Technol. A **24**, 2041 (2006).

[108] E.C. Le Ru, P. Howe, T. Jones, and R. Murray, Phys. Rev. B **67**, 165303 (2003).

[109] O. Flebbe, H. Eisele, T. Kalka, F. Heinrichsdorff, A. Krost, D. Bimberg, and M. Dähne-Prietsch, J. Vac. Sci. Technol. B **17**, 1639 (1999).

[110] J.M. García, J.P. Silveira, and F. Briones, Appl. Phys. Lett. **77**, 409 (2000).

[111] H.J. Scheel, J. Cryst. Growth **211**, 1 (2000).

[112] G.B. Stringfellow, Ann. Rev. Mater. Sci. **8**, 73 (1978).

[113] P.D. Dapkus, Ann. Rev. Mater. Sci. **12**, 243 (1982).

[114] R.L. Moon, J. Cryst. Growth **170**, 1 (1997).

[115] E. Tokumitsu, Y. Kudou, M. Konagai, and K. Takahashi, J. Appl. Phys. **55**, 3163 (1984).

[116] W.T. Tsang, Appl. Phys. Lett. **45**, 1234 (1984).

[117] A.Y. Cho, J. Cryst. Growth **201-202**, 1 (1999).

[118] W.T. Tsang, IEEE J. Quant. Electron. **20**, 1119 (1984).

[119] W.T. Tsang, J. Vac. Sci. Technol. A **2**, 409 (1984).

[120] M.A. Herman and H. Sitter, *Molecular Beam Epitaxy*, 2nd edition, Springer, Berlin, 1996.

[121] K. Ploog, Ann. Rev. Mater. Sci. **11**, 171 (1981).

[122] W.T. Tsang, R.L. Hartman, B. Schwartz, P.E. Fraley, and W.R. Holbrook, Appl. Phys. Lett. **39**, 683 (1981).

[123] S. Hiyamizu, T. Fujii, S. Muto, T. Inata, Y. Nakata, Y. Sugiyama, and S. Sasa, J. Cryst. Growth **81**, 349 (1987).

[124] P.A. Maki, S.C. Palmateer, A.R. Calawa, and B.R. Lee, J. Vac. Sci. Technol. B **4**, 564 (1986).

[125] M. Knudsen, Ann. Phys. **28 (333)**, 999 (1909).

[126] H. Hertz, Ann. Phys. & Chem. **XVII**, 177 (1882).

[127] M. Knudsen, Ann. Phys. **47 (352)**, 697 (1915).

[128] I. Langmuir, Z. Phys. **14**, 1273 (1913).

[129] J.H. Neave, P.K. Larsen, J.F. van der Veen, P.J. Dobson, and B.A. Joyce, Surf. Sci. **133**, 267 (1983).

[130] A. Madhukar, Surf. Sci. **132**, 344 (1983).

[131] M.A. Herman, *Semiconductor Superlattices*, Akademie-Verlag, Berlin, 1986.

[132] B. Lewis and J.C. Anderson, *Nucleation and Growth of Thin Films*, Academic Press, New York, 1978.

[133] R.L. Schwoebel and E.J. Shipsey, J. Appl. Phys. **37**, 3682 (1966).

[134] S. Ino, Jpn. J. Appl. Phys. **16**, 891 (1977).

[135] P.A. Maksym and J.L. Beeby, Surf. Sci. **110**, 423 (1981).

[136] M. Henzler, Surf. Sci. **73**, 240 (1978).

[137] J.J. Harris, B.A. Joyce, and P.J. Dobson, Surf. Sci. Lett. **103**, L90 (1981).

[138] J.H. Neave, B.A. Joyce, P.J. Dobson, and N. Norton, Appl. Phys. A **31**, 1 (1985).

[139] J.M. Van Hove, C.S. Lent, P.R. Pukite, and P.I. Cohen, J. Vac. Sci. Technol. B **1**, 741 (1983).

[140] P.K. Larsen, P.J. Dobson, J.H. Neave, B.A. Joyce, B. Bölger, and J. Zhang, Surf. Sci. **169**, 176 (1986).

[141] W. Faschinger, P. Juza, and H. Sitter, J. Cryst. Growth **115**, 692 (1991).

[142] J. Sudijono, M.D. Johnson, C.W. Snyder, M.B. Elowitz, and B.G. Orr, Phys. Rev. Lett. **69**, 2811 (1992).

[143] P.N. Fawcett, J.H. Neave, J. Zhang, and B.A. Joyce, Surf. Sci. **296**, 67 (1993).

[144] J.P.A. van der Wagt and J.S. Harris Jr., J. Vac. Sci. Technol. B **12**, 1236 (1994).

[145] L.C. Cai, H. Chen, C.L. Bao, Q. Huan, and J.M. Zhou, J. Cryst. Growth **197**, 364 (1999).

[146] Á. Nemcsics and F. Riesz, Cryst. Res. Technol. **36**, 1011 (2001).

[147] S. Martini, A.A. Quivy, T.E. Lamas, and E.C.F. da Silva, Phys. Rev. B **72**, 153304 (2005).

[148] D.W. Pashley, J.H. Neave, and B.A. Joyce, Surf. Sci. Lett. **603**, L1 (2009).

[149] Á. Nemcsics, Ch. Heyn, A. Stemmann, A. Schramm, H. Welsch, and W. Hansen, Mat. Sci. & Eng. B **165**, 118 (2009).

[150] J. Griesche, J. Cryst. Growth **149**, 141 (1995).

[151] W. Braun, *Applied RHEED*, Springer, Berlin, 1999.

[152] S. Noda und A. Sasaki Y. Nabetani, T. Ishikawa, J. Appl. Phys. **76**, 347 (1994).

[153] H. Lee, R. Lowe-Webb, W. Yang, and P.C. Sercel, Appl. Phys. Lett. **72**, 812 (1998).

[154] B.A. Joyce, Rep. Prog. Phys. **48**, 1637 (1985).

[155] J.H. Neave, P.J. Dobson, B.A. Joyce, and J. Zhang, Appl. Phys. Lett. **47**, 100 (1985).

[156] B.A. Joyce, P.J. Dobson, J.H. Neave, K. Woodbridge, J. Zhang, P.K. Larsen, and B. Bölger, Surf. Sci. **168**, 423 (1986).

[157] S. Clarke and D.D. Vvedensky, Phys. Rev. Lett. **58**, 2235 (1987).

[158] G. Binnig, H. Rohrer, Ch. Gerber, and E. Weibel, Phys. Rev. Lett. **49**, 57 (1982).

[159] G. Binnig and H. Rohrer, Helv. Phys. Ac. **55**, 726 (1982).

[160] W. Wu, J.R. Tucker, G.S. Solomon, and J.S. Harris, Appl. Phys. Lett. **71**, 1083 (1997).

[161] R.M. Feenstra and J.A. Stroscio, J. Vac. Sci. Technol. **5**, 923 (1987).

[162] R.M. Feenstra and P. Mårtensson, Phys. Rev. Lett. **61**, 447 (1988).

[163] G. Binnig, C.F. Quate, and Ch. Gerber, Phys. Rev. Lett. **56**, 930 (1986).

[164] D.W. Pohl, W. Denk, and M. Lanz, Appl. Phys. Lett. **44**, 651 (1984).

[165] R.H. Fowler and L. Nordheim, Proc. Roy. Soc. A **119**, 173 (1928).

[166] J. Tersoff and D.R. Hamann, Phys. Rev. Lett. **50**, 1998 (1983).

[167] J. Chen, *Introduction to Scanning Tunneling Microscopy*, Oxford University Press, New York, 1993.

[168] R. Wiesendanger (Ed.), *Scanning Probe Microscopy and Spectroscopy. Analytical Methods*, NanoScience and Technology, Springer, Berlin, 1998.

[169] P. Geng, J. Márquez, L. Geelhaar, J. Platen, C. Setzer, and K. Jacobi, Rev. Sci. Instrum. **71**, 504 (2000).

[170] L. Geelhaar, J. Márquez, K. Jacobi, A. Kley, P. Ruggerone, and M. Scheffler, Microelectron. J. **30**, 393 (1999).

[171] L. Geelhaar, J. Márquez, and K. Jacobi, Phys. Rev. B **62**, 6908 (2000).

[172] Y. Temko, T. Suzuki, M.C. Xu, and K. Jacobi, Appl. Phys. Lett. **83**, 3680 (2003).

[173] Y. Temko, T. Suzuki, P. Kratzer, and K. Jacobi, Phys. Rev. B **68**, 165310 (2003).

[174] T. Van Buuren, M.K. Weilmeier, I. Athwal, K.M. Colbow, J.A. Mackenzie, T. Tiedje, P.C. Wong, and K.A.R. Mitchell, Appl. Phys. Lett. **59**, 464 (1991).

[175] C.T. Foxon and B.A. Joyce, Surf. Sci. **50**, 434 (1975).

[176] C.T. Foxon and B.A. Joyce, Surf. Sci. **64**, 293 (1977).

[177] C.G. Morgan, P. Kratzer, and M. Scheffler, Phys. Rev. Lett. **82**, 4886 (1999).

[178] M. C. Xu, Y. Temko, T. Suzuki, and K. Jacobi, J. Appl. Phys. **98**, 083525 (2005).

[179] J. Grabowski, C. Prohl, B. Höpfner, M. Dähne, and H. Eisele, Appl. Phys. Lett. **95**, 233118 (2009).

[180] E. Penev, P. Kratzer, and M. Scheffler, Phys. Rev. B **64**, 085401 (2001).

[181] E.S. Penev, *On the theory of surface diffusion in InAs/GaAs(001) heteroepitaxy*, PhD thesis, Technische Universität Berlin, 2002.

[182] S. Tsukamoto, T. Honma, G.R. Bell, A. Ishii, and Y. Arakawa, small **2**, 386 (2006).

[183] J. Mirecki Millunchick, A. Riposan, B.J. Dall, C. Pearson, and B.G. Orr, Appl. Phys. Lett. **83**, 1361 (2003).

[184] W. Barvosa-Carter, R.S. Ross, C. Ratsch, F. Grosse, J.H.G. Owen, and J.J. Zinck, Surf. Sci. Lett. **499**, L129 (2002).

[185] P. Kratzer, E. Penev, and M. Scheffler, Appl. Surf. Sci. **216**, 436 (2003).

[186] S. Ohkouchi and A. Gomyo, Appl. Surf. Sci. **130**, 447 (1998).

[187] A. Ishii, K. Fujiwara, and T. Aisaka, Appl. Surf. Sci. **216**, 478 (2003).

[188] H. Eisele, B. Höpfner, C. Prohl, J. Grabowski, and M. Dähne, Surf. Sci. **604**, 283 (2010).

[189] H. Yamaguchi and Y. Horikoshi, Phys. Rev. B **51**, 9836 (1995).

[190] C. Ratsch, W. Barvosa-Carter, F. Grosse, J.H.G. Owen, and J.J. Zinck, Phys. Rev. B **62**, R7719 (2000).

[191] J.M. Márquez Bertoni, *Struktur von GaAs-Oberflächen und ihre Bedeutung für InAs-Quantenpunkte*, PhD thesis, Technische Universität Berlin, 2000.

[192] C. Prohl, B. Höpfner, J. Grabowski, M. Dähne, and H. Eisele, J. Vac. Sci. Technol. B **28**, C5E13 (2010).

[193] R.H. Miwa and G.P. Srivastava, Phys. Rev. B **62**, 15778 (2000).

[194] P. Kratzer, C.G. Morgan, and M. Scheffler, Phys. Rev. B **59**, 15246 (1999).

[195] B. Alloing, C. Zinoni, L.H. Li, A. Fiore, and G. Patriarche, J. Appl. Phys. **101**, 024918 (2007).

Die VDM Verlagsservicegesellschaft sucht für wissenschaftliche Verlage abgeschlossene und herausragende

Dissertationen, Habilitationen, Diplomarbeiten, Master Theses, Magisterarbeiten usw.

für die kostenlose Publikation als Fachbuch.

Sie verfügen über eine Arbeit, die hohen inhaltlichen und formalen Ansprüchen genügt, und haben Interesse an einer honorarvergüteten Publikation?

Dann senden Sie bitte erste Informationen über sich und Ihre Arbeit per Email an *info@vdm-vsg.de*.

Sie erhalten kurzfristig unser Feedback!

VDM Verlagsservicegesellschaft mbH
Dudweiler Landstr. 99   Telefon +49 681 3720 174
D - 66123 Saarbrücken   Fax   +49 681 3720 1749
www.vdm-vsg.de

Die VDM Verlagsservicegesellschaft mbH vertritt

Printed by Books on Demand GmbH, Norderstedt / Germany